POPULAR DOGS' BREED SERIES

ALSATIAN	*Joseph Schwabacher and Thelma Gray*
BASSET HOUND	*George Johnston*
BEAGLE	*Thelma Gray*
BOXER	*Elizabeth Somerfield*
CAIRN TERRIER	*J. W. H. Beynon and Alex Fisher*
CAVALIER KING CHARLES SPANIEL	*Mary Forwood*
CHIHUAHUA	*Thelma Gray*
COCKER SPANIEL	*Veronica Lucas-Lucas*
COLLIE	*Margaret Osborne*
DACHSHUND	*E. Fitch Daglish*
DALMATIAN	*Eleanor Frankling*
FOX TERRIER	*Elsie Williams*
GOLDEN RETRIEVER	*Joan Tudor*
GREYHOUND	*H. Edwards Clarke*
PEKINGESE	*John A. Vlasto and Mary de Pledge*
POODLE	*Clara Bowring and Alida Monro*
PUG	*Susan Graham Weall*
SCOTTISH TERRIER	*Dorothy Caspersz*
SHETLAND SHEEPDOG	*Margaret Osborne*
SPRINGER SPANIEL	*Dorothy Morland Hooper*
WELSH CORGI	*Charles Lister-Kaye*
WHIPPET	*C. H. Douglas-Todd*
YORKSHIRE TERRIER	*Ethel Munday*
WEST HIGHLAND WHITE TERRIER	*D. Mary Dennis*

On the way

GREAT DANE	*Jean Lanning*
STAFFORDSHIRE BULL TERRIER	*John F. Gordon*
AFGHAN HOUND	*Charles Harrisson*
LABRADOR RETRIEVER	*Lorna, Countess Howe and Geoffrey Waring*

THE

FOX TERRIER

(Wire and Smooth)

POPULAR DOGS PUBLISHING CO., LTD
178–202 Great Portland Street, London W1
AN IMPRINT OF THE HUTCHINSON GROUP
London Melbourne Sydney Auckland
Bombay Toronto Johannesburg New York

First published (as *The Popular Fox Terrier*) *1965*
Revised and reprinted (as *The Fox Terrier*) *1970*

This book has been set in Times, printed in Great Britain
on Antique Wove paper by Anchor Press, and
bound by Wm. Brendon both of Tiptree, Essex
09 075032 2

To my long-suffering husband

ACKNOWLEDGMENTS

The two verses from 'How the Camel Got His Hump' from *Just So Stories* by Rudyard Kipling are included by permission of Mrs George Bambridge and Messrs Macmillan & Co. Ltd.

CONTENTS

ILLUSTRATIONS

IN THE TEXT

AUTHOR'S INTRODUCTION

As a little girl I had a passion for animals. I am afraid that my thoughts on these in school hours greatly interfered with my studies! It is significant that I was placed in charge of the girls' pets, including two donkeys!

So when I was invited to rewrite *The Popular Fox Terrier* (the first version by Rev. A. J. Skinner appeared in 1925, and the second by Rev. Dr Rosslyn Bruce in 1950) my immediate reaction was that I was no 'Somerset Maugham' to undertake such a literary task. However, upon reflection, I came to believe that perhaps from my great many years of practical experience of Fox Terriers I might have something to say that could possibly prove of value to the novice breeder and exhibitor, and which might outweigh the obvious limitations of my literary abilities. I was encouraged in this belief by the many requests I have received from my friends for tips and advice which, they exclaimed, 'should fill a book'.

So I have undertaken the work, and at last it is over. Had I known of the amount of research involved, and the toils and tribulation of literary creation, I very much doubt whether I should ever have started. However, I have been greatly assisted in the work by my friends. In particular I must record my gratitude to Miss Kathleen Emery, the President of the Fox Terrier Club, who approved my chapter on 'Smooth Fox Terriers, dominant and famous sires and dams'; to Mr B. S. Jordan, Hon. Secretary of the Wire Fox Terrier Association, and to Mr John Lowe, Hon. Secretary of the Smooth Fox Terrier Association, both of whom gave me permission to use matter from their respective year books. In addition, I must thank the officers of the Kennel Club for permission to quote the standards of Wire Fox Terriers and Smooth Fox Terriers, and express my appreciation of the remarkably complete records kept by the Kennel Club, which have enabled me to compile the schedules of pedigrees of post-war champions.

Finally, I am very grateful to the many friends who have loaned to me photographs of their terriers for reproduction in this book. I hope that they will not be disappointed by what I have to say about their 'stars'!

Fox Terriers have been my life's hobby. My enthusiasm for them should be apparent. If what I have written brings to my readers even a small part of the pleasure and reward which I have enjoyed from my dogs, I shall count my book as well worth while.

1965 E.L.W.

From the many letters I have received from all parts of the world I gather that my efforts in writing this book have been appreciated and have given encouragement and help to many. I hope, too, that it has imparted a little of my enthusiasm and hope for the future to all who have read it.

It is gratifying that the demand for the book has been sufficient to justify this second edition, now published under the title of *The Fox Terrier*.

I have brought up to date the breeding of our latest Champions, both Wire and Smooth, and also the names and addresses of the secretaries of the various Fox Terrier Clubs; registrations of both coats 1965–1968 have been added. A few minor revisions have also been made to the text, and two of the photographs have been replaced.

1970 E.L.W.

CHAPTER I

ORIGINS AND EVOLUTION

TERRIERS have always been popular. Their convenient size, gameness and gay character have appealed throughout the ages. Their origin is obscure, but they probably evolved as a type from man's need of short-legged dogs able to go down holes in the earth after game or vermin. In consequence, man continually bred from dogs with short legs, and so a short-legged strain was established which bred more or less to type. Such a dog is recorded as early as A.D. 1500, but the short-legged dog must have been used and valued centuries before this.

From the short-legged strain emerged local types, both rough-coated and smooth, which by a process of selective breeding have been channelled into terriers of individual characteristics which are now recognized by the Kennel Club as distinct breeds.

The Rev. Dr Rosslyn Bruce, in an earlier edition of *The Popular Fox Terrier* (1950), records that: 'The earliest writers include Dr Caius (1590); and Marco Polo, who tells us that in the thirteenth century the Grand Khan had 5,000 various "hounds"; besides the Chinese, the Egyptian empresses made pets chosen from out of their lords' kennels; and later our Roman conquerors had little shooting dogs too small to fight in the arena. In 1486 our own Dame Juliana Berners alludes, though sparingly, to "terroures" in her *Book of St Albans*.

> Body and limb go cold, go cold,
> Both foot and hand go bare;
> God send terroures so bold, so bold,
> Heart will harbour no care:

playfully wrote Dr Still, Bishop of Bath, about 1570. Dr Caius was a Cambridge professor about 1580, and he wrote the first book on

Englishe Dogges in Latin; he gives details of the work of a "Terrarius" with fox and badger, and compares him to a ferret! Gervase Markham wrote sketchily about terriers in 1600; Nicholas Cox in *The Gentleman's Recreation* in 1667 told of two varieties of small dogs, who go to ground after game! Many early French writers, too, spoke about terriers at this period. In 1718 in the *Compleat Sportsman* Nicholas Cox describes the work of a terrier, without considering his appearance or colour. The first distinction between rough and smooth terriers appears to come from Daniel in *Field Sports* about 1760 and he also deals with their colour.

'Probably the first record of painting of a Fox Terrier of the modern type, both in colour and shape, is one seen in Holland about 1880 by Mr (later, Sir) J. A. Doyle, an Oxford don of very distinguished reputation; this picture had a background of fruit and flowers, and was painted by a Dutch artist called Hamilton, about 1700. Mr Rawdon Lee, writing about 1890 in his memorable volume *The Fox Terrier* (which ran into numerous editions, of which the 4th is dated 1902), states that he has a copy of an old engraving before him as he writes of "King James I, hawking"; at the feet of the monarch are four dogs, evidently terriers, one of which is almost, if not quite, white, with a well-shaped terrier-like head: then again we have the ugly (but now famous) picture, painted by Sartorius in 1796, of Viper, a white dog with clear black patches, discovered by Capt. Keene shortly before he died in 1896: another picture by Sartorius shows two terriers, one all white, and a second with two black patches and a little tan on the right eye-mark.'

The original short-legged dog was probably a black and tan with a patch of white. It would be a game dog, quick with rats and eager for battle in fox or badger earth. It would be small but sturdy, tough, with much endurance; able to keep up with the huntsmen and to fight to the death. Such qualities are evident today in the modern terrier, both working and show strains.

My very first Wire Fox Terrier was purchased forty years or more ago from Her Grace the late Duchess of Newcastle for the sum of £1. He was an eight-week-old dog puppy by Round-up ex Wycollar Girl. This dog was subsequently registered as Grey Beard Loon, shown, and, to my great delight, gained show prizes. He and his daughter Puck bolted a badger in broad daylight from among the rocks of Clodgy, near St Ives, in Cornwall. As for fighting, woe betide any dog, however large, who tackled either of them, as the other would at once join in, and together they would rout the largest assailant.

Fox Terriers probably date as a type from the late seventeenth or early eighteenth century, when foxhunting became popular as a sport. Formerly the quarry had been deer or hare, which remained above ground. The fox, it was found, went to earth and had to be bolted if the hunt were to continue. Terriers were therefore in demand, especially those with plenty of white in their coats. The black-and-tan terriers were too frequently mistaken for the fox when emerging from the earth and were set upon by the hounds. A white strain of terrier was consequently bred, about which there could be no mistake in this respect.

The terrier had to be small enough to penetrate the fox earth. At the same time it had to have qualities of running power and endurance to enable it to follow hounds. The terrier with very short legs failed in the latter respect. It is therefore probable that selected terriers were crossed with Foxhounds or Beagles to produce smooth-coated dogs of greater length of leg but with terrier qualities of character. Some authorities contend that the white strain of the Bull Terrier was introduced to give the desirable white predominance. By selective breeding, a type closely resembling the present-day Smooth Fox Terrier, with white predominating, was stamped-in and came to be regarded as the typical and desirable Fox Terrier. It is interesting to note that the modern Kennel Club standard of the Fox Terrier, of both coats, stipulates that white must predominate. In this the standard perpetuates the needs of the foxhunter. Not all the early Fox Terriers had smooth coats. Some retained traces, in varying strength, of the rough, wiry coats of the original terrier-like dogs from which they sprang. These were not favoured, but the rough strain persisted.

By the beginning of the nineteenth century the Fox Terrier had become established as a special type or breed. However, it was not until 1876 that a number of breeders met together, at the suggestion of the late Major Harding Cox, to form the Fox Terrier Club for the purpose of improving and maintaining the strain or breed. The smooth and wire coats were recognized as different species of the same family. Nevertheless, the smooth-coated terrier remained the more popular. At the first major show for terriers in 1886, at the

Royal Aquarium in Westminster, of which the late Mr Charles Cruft was manager, 125 Smooths were exhibited as against fifty wire-haired terriers.

This trend did not continue, at least among show folk. By 1888 both Smooths and Wires were in great demand. In his book *This Doggie Business* Mr Edward C. Ash records that 'although in 1901 smooth Fox Terriers were still holding their own, the Wire terriers, once despised, were leaving the Smooths to play a back number. That this should happen is mainly due to discoveries in the way of preparing the terrier for shows, for if properly prepared, the breed is most attractive, even more attractive than the smooth.'

As a family pet, however, the smooth variety gained great popularity, perhaps because their coats were so much easier to keep in condition than the wire breed.

As evidence of the demand for Fox Terriers in the early days, Mr Ash quotes some astonishing sale prices: 'At the Fourth Terrier Show Mr Wootton, acting for Lord Lonsdale, bought the wire-coated terrier Briggs for £200; Miss Nuggs for £105; Vera for £105; Sam Weller for £100; Snowball for £100 and Bundle for £42. Such prices for "outcasts" came as an awful shock to smooth-coated breeders.'

Later, Mr Ash mentions the sale by Mr George Raper of Champion Go Bang for £500 to the United States. This dog was one of the best Wire terriers at the Birmingham Show in 1898. Mr Raper also purchased the Duchess of Newcastle's Coastguard of Notts for £150 and, by public auction, Matchmaker for £250.

When one considers the relative value of the pound then, compared with the present day, these prices seem astonishing. There have been several subsequent sales of terriers, both smooth-coated and wire, for sums of four figures, but I feel that the sales quoted in the early days must be an indication of the high regard which dog fanciers of the time had for the sporting qualities and attributes of character which the breed demonstrated.

Alongside with the development of the Fox Terrier for show purposes, the working terriers kept for foxhunting, badger-baiting and ratting had evolved into a traditional type, both smooth- and rough-coated, somewhat shorter in the leg. Parson Jack Russell of Devonshire gained fame for his strain in the mid-nineteenth century. His terriers followed hounds and were consequently higher on the leg than later became the fashion for working purposes, when it was

Viper, 1796

Ch. Talavera Simon
(Born 1924)

Hedges

Ch. Beau Brummel
(Born 1928)

Metcalfe

Ch. Galant Fox of Wildoaks
(Born 1929)

Ch. Weltona Revelation
(Born 1943)

Thurse

Thurse

Ch. Weltona Exelwyre Dustynight
(Born 1947)

N. H. J. Baird's frontispiece to the 1902 limited edition of *Memoir of the Rev. John Russell* by E. W. L. Davies the custom for hunt terriers to be carried by the mounted terrier man. From contemporary drawings of the Parson's terriers it is apparent that the so-called Jack Russell terriers of the present time are quite unlike the Sporting Parson's terriers. The modern working terrier is very short-legged, but equally game and gay.

In 1899 *The Stable*, a weekly journal, described the character of the Fox Terrier in the following terms:

'No one will dispute the right of the Fox Terrier to figure first in the ranks of popular breeds. Attractive in appearance, cheerful in disposition, ever alert, and always displaying his powers, he is the beau ideal of a canine companion. Take him into the show ring, and the fire and energy of the breed at once become apparent, while in the family circle he is always the prime favourite, and leading artists constantly depict how popular the Fox Terrier has become.'

I cannot improve upon this description. My whole experience of the breed, spread over many years, has shown it to be true. Countless incidents of loyalty, gameness, playful ways and devotion have led to my love of the Fox Terrier from which I have never swerved. Among family pets I recall Jester, the badger-baiter, who delighted to ride with me on my bicycle with his front feet upon the handlebars, and of his habit of climbing ladders on to straw-ricks; of Lady's devotion and anxiety during my absence; of Puck's way of throwing rats over her head; of Pomp's zeal as my personal guard; of Happy's eagerness for an outing; of Deborah's preference for a *red* ball, and of Peanuts' ability to distinguish instantly between the family cat and the ginger tom from next door, and of his interest in my husband's fishing activities.

All these qualities endear the breed to me, and I would never wish to be without a Fox Terrier as a companion or showman. While they have brought to me a great deal of work, I have gained a lot of happiness from their company. I have every intention that they shall continue to do so.

I do not think I can do better than quote *The Fox Terrier's Alphabet* as given by the late Rev. Dr Rosslyn Bruce:

'A for my Action, straight, spritely, and sound,
B for my Bone, which is ample and round;
C for my Coat, Smooth, or Wire, but hard,
D for Distemper, which I'm bred to discard;

E for my Ears and Eyes, lively and small,
F for my Feet, hardly showing at all;
G for my Gameness with fox or with vermin,
H for my Head, which my parents determine;
I for In-breeding, since my breeding is sound,
J for my Jaw for use underground;
K for my Kennelman, skilful and handy,
L for my Legs, neither knock-kneed nor bandy;
M for my Muscles on quarters and back,
N for my Nose, which is keen, and is black;
O for my Owner, who glories in winners,
P for the Prize-cards, which comfort beginners;
Q for the Questions concerning our merits,
R for Replies from the rats and the ferrets;
S for my Shoulders, fine, sloping, and long,
T for my Teeth, close fitting and strong;
U for the Use, which I prove in your sport,
V for the Value "of such a good sort";
W for White, which should always predominate;
Y for my Youth and its happy unrest,
Z for your Zeal for me, British and Best.'

From the appearance of the Fox Terrier in the show ring in the latter half of the last century the breed gained in popularity until a peak registration with the Kennel Club of 7,156 Wires and 2,573 Smooths was reached in 1924. Today the Fox Terrier has given place numerically to the Poodle and other more fashionable breeds, but no one can deny that the quality and character of the modern Fox Terrier is as attractive as ever. Registrations, which, during the first few years after the Second World War, amounted to about 6,500 annually for Wires and 2,000 for Smooths, have now settled down to an average of 2,300 and 750 respectively. Some may deplore the reduction in numbers, but I, for one, much prefer the maintenance of quality to quantity. The fact that the interest in the breed remains as high as ever is evidenced by the throngs of knowledgeable spectators round the Fox Terrier rings at any championship show. This must be a source of satisfaction to anyone who has the welfare and interest of Fox Terriers at heart.

CHAPTER 2

DESCRIPTION AND STANDARDS

THE parent club of Fox Terriers, formed in 1876 in London, drew up a descriptive standard both for smooth and wire coats. These are practically the same today, and are published each year in the members' year books by the three societies: namely, The Fox Terrier Club, the Wire Fox Terrier Association and the Smooth Fox Terrier Association, after approval by the Kennel Club. The object of these clubs is to help the breed in every way possible and to support shows up and down the country. All three societies run their own championship shows every year, where their very valuable cups are offered to members. The shows are open to all, and a win at one of these shows is considered a great honour. The standard is issued not as a hard-and-fast rule for breeders and judges but more as a guide. No two people see dogs alike, and each judge will forgive one fault more than another: for instance, some insist on the terrier having a very long, lean head, while others loathe and detest a long back; yet others demand particularly good, strong quarters, some are very conscious about size, either too small or too large, so it is very easy to understand the saying 'Many men, many minds'.

The clubs also issue:

THE POINTS OF A TERRIER

		Smooth-haired		*Wire-haired*	
1.	Head and ears	15		15	
2.	Neck	5		5	
3.	Shoulders and chest	10		10	
4.	Back and loins	10		10	
5.	Hind quarters	15	100	15	100
6.	Stern	5		5	
7.	Legs and feet	15		15	
8.	Coat	10		15	
9.	Symmetry, size and character	15		10	

I, personally, think the points of little value either in judging or in breeding, as, after all, each dog must be taken as a *whole*, not in separate little pieces. A dog with a long, lean head, long neck let into good shoulders and with good quarters and set-on, lovely legs and feet, good body, perfect shoulders and covered with a good coat, can still not be an outstanding specimen, simply because the parts do not join up into the whole pleasantly to give a good outline with style and character.

Fig. 1 illustrates the different parts of the terrier's body, and should be studied by beginners in order that they can understand the details of the standard given below.

THE STANDARD OF THE WIRE FOX TERRIER

Reproduced by permission of the Kennel Club

CHARACTERISTICS *The Terrier should be alert, quick of movement, keen of expression, on the tip-toe of expectation at the slightest provocation. Character is imparted by the expression of the eyes and by the carriage of ears and tail.*

This does not mean that a 'jumpy' highly strung terrier is required. Nothing is worse than one that is completely unmanageable either in the house, kennel or the show ring. They should certainly be alert and active, and—and this is most important—manageable. You may have a certain champion in the house or kennel where everything proceeds in an even rhythm, but if when you take him out for a walk he pulls and tugs on the lead for all he is worth, and when he sees another dog almost pulls your arm out of its socket, this is most undesirable and takes away all pleasure when exercising him. It also spoils his chances in the show ring. In my opinion, a dog must not only be a good specimen of the breed, but he must also be manageable, and while being shown he must be an actor, look the part, and be an active, gay, alert and controllable dog under all circumstances.

GENERAL APPEARANCE *The dog should be balanced and this may be defined as the correct proportions of a certain point or points, when considered in relation to a certain other point or points. It is the keystone of the Terrier's anatomy. The chief points for consideration are the relative proportions of skull and foreface; head and back; height at withers; and length of body from shoulder-point to buttock—the ideal of proportion being reached when the last two measurements are the same. It should be added that, although the head measurements can be taken with absolute accuracy, the height at withers and length of back are approximate, and are inserted for the information of breeders and exhibitors rather than as a hard-and-fast rule. The movement or action is the crucial test of conformation. The Terrier's legs should be carried straight forward while travelling, the forelegs hanging perpendicular and swinging parallel to the sides, like the pendulum of a clock. The principal propulsive power is furnished by the hind legs, perfection of action being found in the Terrier possessing long thighs and muscular second-thighs well bent at the stifles, which admit of a strong forward thrust or 'snatch' of the hocks. When approaching, the forelegs should form a continuation of the straight of the front, the feet being the same distance apart as the elbows. When stationary it is often difficult to determine whether a dog is slightly out at shoulder but directly he moves the defect—if it exists—becomes more apparent, the fore-feet having a tendency to cross, 'weave' or 'dish'. When, on the contrary, the dog is tied at the shoulder, the tendency of the feet is to move wider apart, with a sort of padding action. When the hocks are turned in—cow hocks—the stifles and feet are turned outwards, resulting in a serious loss of propulsive power. When the hocks are turned outwards the tendency of the hind feet is to cross, resulting in an ungainly waddle.*

Much is written about balance in a terrier. I have argued, on many occasions, that balance is not quite the correct word to use. Outline, to me, is more correct. Balance means, surely, that everything must be in proportion. I find that to win, one must have slight exaggerations. For instance, just a little longer, leaner head, just a little bit shorter in back, just a little more bone, just a little better set-on, just a little longer neck—*provided* the whole makes up into a good outline—must surely win over a so-called balanced dog or bitch.

As the standard says, good movement is the crucial test of correct conformation. Narrow chests and tied shoulders, which give the impression of the two front legs coming out of the one hole, show very quickly, for as soon as the dog walks he crosses his front legs over each other in a 'Tishy' action instead of striding forward with good, long strides. Short shoulder blades, even with a good long neck, make the dog look as if he is 'reaching for the ground', almost

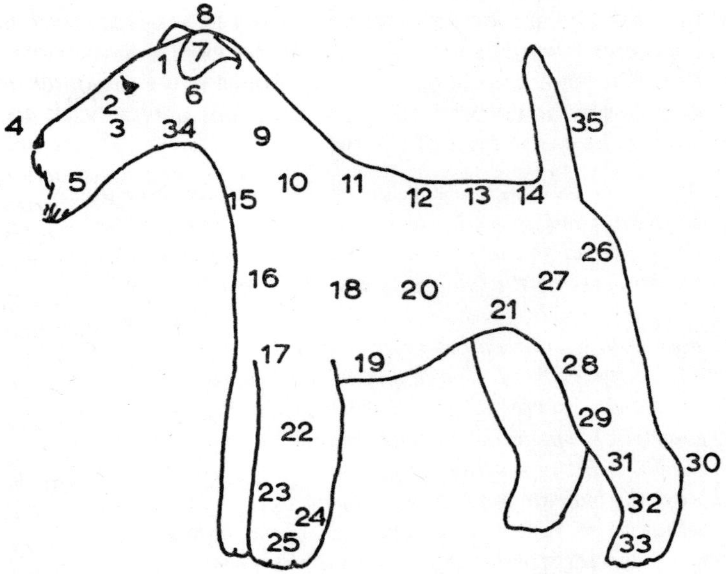

Wire Fox Terrier Association

Fig. 1 Points of the Terrier

1 Forehead
2 Junction of Forehead and Nose
3 Nose (Muzzle)
4 Nostrils
5 Muzzle (proper)
6 Temple
7 Ears
8 Occiput
9 Posterior Angle of Jaw
10 Neck
11 Withers
12 Saddle
13 Loins
14 Croup
15 Dewlap
16 Brisket
17 Arm
18 Shoulder
19 Point of Elbow
20 Ribs and Chest Wall
21 Flank
22 Forearm
23 Knee or Waist
24 Pastern
25 Toes
26 Buttock
27 First Thigh
28 Stifle
29 Second Thigh
30 Point of Hock
31 Front Face of Hock
32 Pastern
33 Toes
34 Cheeks
35 Stern

as if his front legs are too short for him. When he is out at elbow he turns his front inwards and even, at times, looks pigeon-toed when standing. Although it is not expressly mentioned in the standard, the foot as a whole should bend neither inwards nor outwards, though the latter is the worse fault of the two.

HEAD AND SKULL *The top line of the skull should be almost flat, sloping slightly and gradually decreasing in width towards the eyes, and should not exceed 3½ inches in diameter at the widest part—measuring with the calipers—in the full-grown dog of correct size, the bitch's skull being proportionately narrower. If this measurement is exceeded the skull is termed 'coarse', while a full-grown dog with a much narrower skull is termed 'bitchy' in head. The length of the head of a full-grown well-developed dog of correct size—measured with calipers—from the back of the occipital bone to the nostrils—should be from 7 to 7¼ inches, the bitch's head being proportionately shorter. Any measurement in excess of this usually indicates an over-sized or long-backed specimen, although occasionally—so rarely as to partake of the nature of a freak—a Terrier of correct size may boast a head 7½ inches in length. In a well-balanced head there should be little apparent difference in length between skull and foreface. If, however, the foreface is noticeably shorter, it amounts to a fault, the head looking weak and 'unfinished'. On the other hand, when the eyes are set too high up in the skull, and too near the ears, it also amounts to a fault, the head being said to have a 'foreign appearance'. Although the foreface should gradually taper from eye to muzzle and should dip slightly at its juncture with the forehead, it should not 'dish' or fall away quickly below the eyes, where it should be full and well made up, but relieved from 'wedginess' by a little delicate chiselling. While well-developed jaw-bones, armed with a set of strong, white teeth, impart that appearance of strength to the foreface which is so desirable, an excessive bony or muscular development of the jaws is both unnecessary and unsightly, as it is partly responsible for the full and rounded contour of the cheeks to which the term 'cheeky' is applied. Nose should be black.*

Here, the standard mentions 'cheeky', as applied to the head, and another of the disqualifying points in the show ring is 'much undershot or much overshot'. An undershot or overshot dog is illustrated, in Fig. 2. For 'cheekiness', see Fig. 3.

One seldom sees a nose of any other colour than black but,

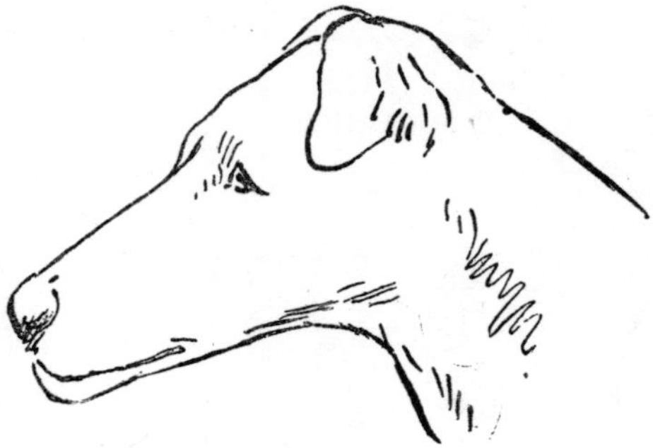

Undershot.

Overshot.

Fig. 2

occasionally, a small strip down the front of the nose will fade a little to a browny colour—this, I find, usually turns black again during the warm weather.

A 'dish' face is most undesirable and ugly. 'Dish' means that the skull is higher than the muzzle, often with a decided stop and a lumpy head (Fig. 4). The skull should be flat when viewed from the side.

EYES *Should be dark in colour, moderately small and not prominent, full of fire, life, and intelligence; as nearly as possible, circular in shape and not too far apart. Anything approaching a yellow eye is most objectionable.*

EARS *Should be small and V-shaped and of moderate thickness, the flaps neatly folded over and dropping forward close to the cheeks. The top line of the folded ear should be well above the level of the skull. A pendulous ear, hanging dead by the side of the head like a hound's, is*

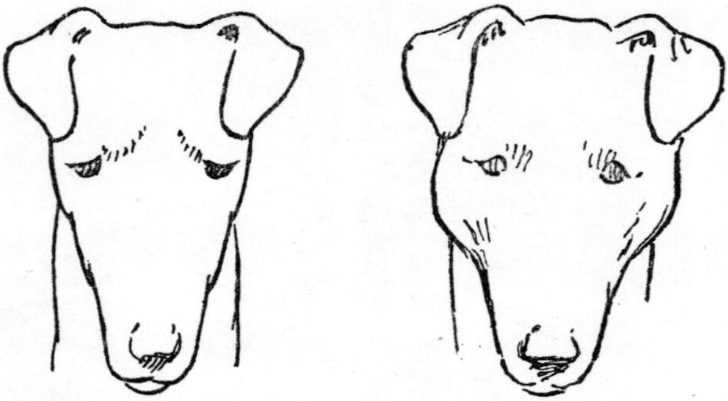

Good head and skull. 'Cheekiness'.

Fig. 3

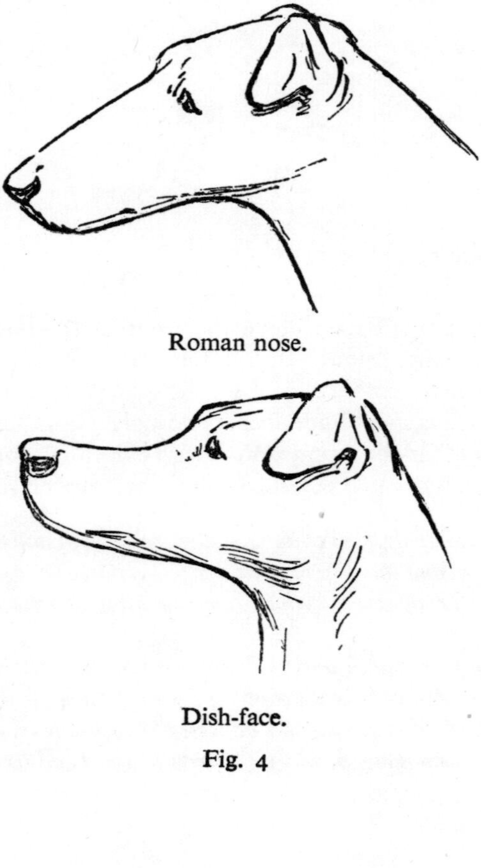

Roman nose.

Dish-face.

Fig. 4

uncharacteristic of the Terrier, while an ear which is semi-erect is still more undesirable.

The position and the liveliness of the ears are most important. Dead, heavy ears take a great deal away from the expression and quality of the head as a whole, and also the general character of the terrier. Lively ears are small, well-placed ears, with the fold above the level of the skull, and ears which the terrier is able to use. By his ears one can judge quite a lot about the mood of a dog. If he is scared or apprehensive he will often screw them back, and if not feeling on top of the world the ears lose the lift and, in consequence, look dull. Quite often in the show ring the judge will drop something in front of the dog to see if he is interested and able to 'use his ears'; that is, put them forward, lift them slightly and thus look really alert and ready for anything.

MOUTH *Both upper and lower jaws should be strong and muscular, the teeth as nearly as possible level and capable of closing together like a vice—the lower canines locking in front of the upper and the points of the upper incisors slightly overlapping the lower.*

Undershot is the term applied when the lower jaw projects beyond the upper one; overshot, when the upper jaw projects beyond the lower. A study of the profile will nearly always reveal if the dog is undershot or not. To make absolutely sure, examine the teeth carefully. The mouth should, of course, be perfectly level: that is, the upper teeth should just close over the under teeth, like the well-fitting lid of an airtight box, though most judges do not penalize a terrier unless either of the above faults is carried to excess.

NECK *Should be clean, muscular, of fair length, free from throatiness and presenting a graceful curve when viewed from the side.*

It is very important that the line from the neck into the shoulders should appear unbroken; there should be no break at the point of the shoulder blades (withers), as this would completely spoil the outline of the terrier. The neck should 'flow' from behind the head, at the back of the ears right into the back line.

FORE QUARTERS *Shoulders when viewed from the front should slope steeply downwards from their juncture, with the neck towards the points, which should be fine. When viewed from the side they should be long, well laid back, and should slope obliquely backwards from points to withers, which should always be clean cut. A shoulder well laid back gives the long fore-hand, which, in combination with a short back, is so desirable in Terrier or Hunter. Chest deep and not broad, a too narrow chest being almost as undesirable as a very broad one. Excessive depth of chest and brisket is an impediment to a Terrier when going to ground Viewed from any direction the legs should be straight, the bone of the fore-legs strong right down to the feet. The elbows should hang perpendicular to the body, working free of the sides, carried straight through in travelling.*

BODY *The back should be short and level, with no appearance of slackness—the loins muscular and very slightly arched. The brisket should be deep, the front ribs moderately arched, and the back ribs deep, and well sprung. The term 'slackness' is applied both to the portion of the back immediately behind the withers when it shows any tendency to dip, and also the flanks when there is too much space between the back-ribs and hip-bone. When there is little space between the ribs and hips, the dog is said to be 'short in couplings', 'short-coupled', or 'well-ribbed-up'. A terrier can scarcely be too short in back, provided he has sufficient length of neck and liberty of movement. The bitch may be slightly longer in couplings than the dog.*

I like a very short back, a long neck and good shoulders, but this is one of the most difficult things of all to obtain. Short-backed dogs often incline to cloddiness and lack freedom of movement.

HIND QUARTERS *Should be strong and muscular, quite free from droop or crouch; the thighs long and powerful; the stifles well curved and turned neither in nor out; the hock-joints well bent and near the ground; the hocks perfectly upright and parallel with each other when viewed from behind. The worst possible form of hind quarters consists of a short second-thigh and a straight stifle, a combination which causes the hind-legs to act as props rather than instruments of propulsion. The hind-legs should be carried straight through in travelling.*

This, is a very good description of correct quarters, and among the ilustrations are various types of hind quarters, including the most desirable (see Fig. 5).

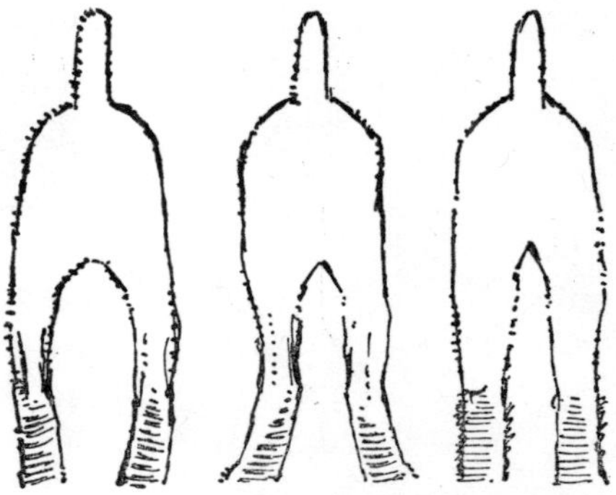

Wide. Cow-hocked. Desirable.

Fig. 5 Hindquarters

FEET *Should be round, compact, and not large—the pads tough and well cushioned, and the toes moderately arched and turned neither in nor out. A Terrier with good-shaped fore-legs and feet will wear his nails down short by contact with the road surface, the weight of the body being evenly distributed between the toe-pads and the heels.*

If terriers do not have sufficient exercise on a hard surface their toenails become very long and will need attention. If the toenails are not kept short the foot becomes open, instead of round and compact.

TAIL *Should be set on rather high and carried gaily but not curled. It should be of good strength and substance and of fair length—a three-quarters dock is about right—since it affords the only safe grip when handling working Terriers. A very short tail is suitable neither for work nor show.*

I prefer to have 'some dog behind the tail': in other words, a well-developed sit-upon and the tail set high. A thick tail usually denotes good bone. An old pig-dealing friend of mine told me years ago that if you choose an animal with a good, strong tail it will have bone and substance. This I have found very true in terriers (see Fig. 6).

COAT *The principal difference between that of the Smooth and Wire variety is that, whereas the former is straight and flat, that of the latter*

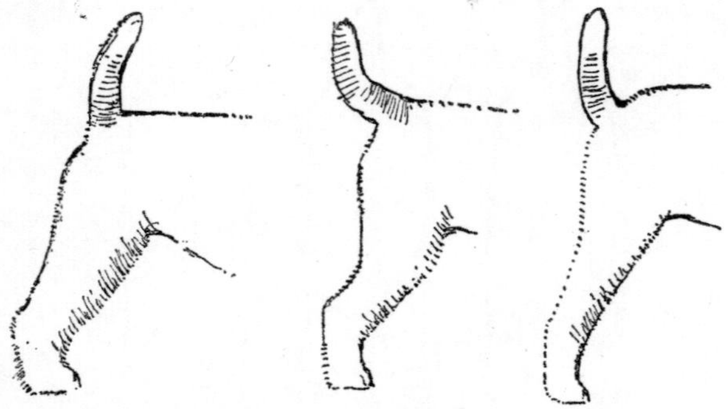

Good set-on. Go-away. Low-set and Straight Stifle.

Fig. 6 Tails

appears to be broken—the hairs having a tendency to twist. The best coats are of a dense, wiry texture—like coconut matting—the hairs growing so closely and strongly together that when parted with the fingers the skin cannot be seen. At the base of these stiff hairs is a shorter growth of finer and softer hair—termed the undercoat. The coat on the sides is never quite so hard as that on the back and quarters. Some of the hardest coats are 'crinkly' or slightly waved, but a curly coat is very objectionable. The hair on the upper and lower jaws should be crisp and only sufficiently long to impart an appearance of strength to the fore-face, thus effectually differentiating them from the Smooth variety. The hair on the forelegs should also be dense and crisp. The coat should average in length from ¾ to 1 inch on shoulders and neck, lengthening to 1½ inches on withers, backs, ribs and quarters. These measurements are given rather as a guide to exhibitors than as an infallible rule, since the length of coat varies in different specimens and seasons. The judge must form his own opinion as to what constitutes a 'sufficient' coat.

The coat must have a hard outer coat and a weather-resisting undercoat. It must never be allowed to become silky or soft, which happens when the undercoat is allowed to grow through the hard outer one.

COLOUR *White should predominate; brindle, red, liver or slaty blue are objectionable. Otherwise, colour is of little or no importance.*

I like an all-white dog, provided it is absolutely white, with no ticking, but most people prefer them hound-marked: that is, an even, tan head, with a white muzzle and blaze, black patches on the body and a little tan perhaps on the tail and quarters. Some are very heavily marked. I well remember a particularly heavily marked bitch winning very high honours, and when the owner was told that according to the standard 'white should predominate' he replied that white *did* predominate if she were turned over to show her tummy! Smooths are preferred with either rich tan markings or black markings on head or body. All-white ones, too, are popular, or with a little marking on the heads. Dirty markings in smooths are not so popular. Still, a great many people think and quote that 'a good horse is never a bad colour'.

WEIGHT AND SIZE *Bone and strength in a small compass are essential, but this must not be taken to mean that a Terrier should be 'cloddy', or in any way coarse—speed and endurance being requisite as well as power. The Terrier must on no account be leggy, nor must he be too short on the leg. He should stand like a cleverly-made, short-backed Hunter, covering a lot of ground. According to present-day requirements, a full-sized, well-balanced dog should not exceed 15½ inches at the withers—the bitch being proportionately lower—nor should the length of back from withers to root of tail exceed 12 inches, while to maintain the relative proportions, the head—as before mentioned—should not exceed 7½ inches or be less than 7 inches. A dog with these measurements should scale 18 lb. in show condition—a bitch weighing some 2 lb. less—with a margin of 1 lb. either way.*

I was taught to believe that a too-big wire, while it could quite often prove to be a good stud proposition, is not desirable for show. A correct-sized terrier is an absolute must for me. Some people say that 'a good big one will always beat a good little one'. I do not think this is true. In my opinion many terriers shown today are far too big, tall on the leg and long in the back. Of course, it is easier to breed a long head on a big dog, and quite often I am told about wonderful puppies with great long heads. When I ask how tall he measures at the withers, however, I find that he is far too big for his age.

I cannot, myself, tell how big a dog is by guessing his weight. One needs to trust one's eyes, because a dog is as big or as small as he *looks*, not what he weighs or measures. It is equally undesirable to show dogs that are too small, especially if they are also at all 'bitchy' in appearance.

FAULTS *Nose: white, cherry, or spotted to a considerable extent with either of these colours. Ears: prick, tulip, or rose. Mouth: much undershot or much overshot.*
N.B.—*Old scars or injuries, the result of work or accident, should not be allowed to prejudice a Terrier's chance in the show ring, unless they interfere with its movement or with its utility for work or stud.*

The standard and description of the Smooth variety of Fox Terrier is almost exactly that of the Wire, the main difference being the coat. Whereas the coat of the Wire should be as hard and wiry as possible, that of the Smooth should be flat, smooth, hard, dense and abundant. It is interesting to note that, in the scale of points, ten points are suggested for the coat of the Smooth and fifteen points for that of the Wire terrier.

THE STANDARD OF THE SMOOTH FOX TERRIER
Reproduced by permission of the Kennel Club

GENERAL APPEARANCE *The dog must present a general gay, lively and active appearance; bone and strength in a small compass are essentials, but this must not be taken to mean that a Fox Terrier should be cloggy, or in any way coarse; speed and endurance must be looked to as well as power, and the symmetry of the Foxhound taken as a model. The Terrier, like the Hound, must on no account be leggy, nor must he be too short in the leg. He should stand like a cleverly made Hunter, covering a lot of ground, yet with a short back. He will then attain the highest degree of propelling power, together with the greatest length of stride that is compatible with the length of his body.*

HEAD AND SKULL *The skull should be flat and moderately narrow, and gradually decreasing in width to the eyes. Not much 'stop' should be apparent, but there should be more dip in the profile between the forehead and the top jaw than is seen in the case of the greyhound. The cheeks must not be full. The jaw, upper and under, should be strong and muscular, should be of fair punishing strength, but not so in any way to resemble the Greyhound. There should not be much falling away below the eyes. This part of the head should, however, be*

The Totteridge XI, 1897

Thurse

Ch. Lanneau Jewel (Born 1947)

Ch. Hewshott Jaguar (Born 1953)

Thurse

moderately chiselled out, so as not to go down in a straight line like a wedge. The nose, towards which the muzzle must gradually taper, should be black.

EYES *Should be dark in colour, small, and rather deep set, full of fire, life, and intelligence; as nearly as possible circular in shape.*

EARS *Should be V-shaped and small, of moderate thickness, and dropping forward close to the cheek, not hanging by the side of the head like a Foxhound's.*

MOUTH *The teeth should be nearly as possible level, i.e., the upper teeth on the outside of the lower teeth.*

NECK *Should be clean and muscular, without throatiness, of fair length, and gradually widening to the shoulders.*

FORE QUARTERS *The shoulders should be long and sloping, well laid back, fine at the points, and clearly cut at the withers.*

BODY *Chest deep and not too broad. Back should be short, straight and strong, with no appearance of slackness. Loin should be powerful and very slightly arched. The fore ribs should be moderately arched, the back ribs deep; and the dog should be well ribbed up.*

HIND QUARTERS *Should be strong and muscular, quite free from droop or crouch; the thighs long and powerful; hocks near the ground, the dog standing well up on them like a Foxhound, and not straight in stifle.*

FEET *Should be round, compact, and not large. The soles hard and tough. The toes moderately arched, and turned neither in nor out.*

TAIL *Should be set on rather high, and carried gaily, but not over the back nor curled. It should be of good strength.*

COAT *Should be straight, flat, smooth, hard, dense and abundant. The belly and under side of the thighs should not be bare.*

COLOUR *White should predominate; brindle, red or liver markings are objectionable. Otherwise this point is of little or no importance.*

WEIGHT AND SIZE *Weight is not a certain criterion of a Terrier's fitness for his work—general shape, size, and contour are the main points—and if a dog can gallop and stay, and follow his fox up a drain, it matters little what his weight is to a pound or so, though, roughly speaking, 15 to 17 lb. for a bitch and 16 to 18 lb. for a dog in Show condition are appropriate weights.*

FAULTS *Nose white: cherry, or spotted to a considerable extent with either of these colours. Ears: prick, tulip or rose. Mouth: much undershot or much overshot.*

Many years ago the Smooth Fox Terrier Association compiled a 'Little Standard', which they believed contained the nine points of special importance in a Fox Terrier. They are as follows:

THE LITTLE STANDARD

A Smooth Fox Terrier should be active and strong, with a brave, wise way with him, to suit him for the fields or home. His head should be long and lean, with much strength in front of his eyes.

His eyes should be small and dark, and full of the pride of life.

His ears should be small and like a 'V' set high on his head.

His neck should be long and arched, his shoulders should slope back, and should also be long.

His fore legs should be straight, round, and thick; his hind legs strong but reachy, and his hocks should be near the ground.

His feet should be small and round, with thick pads.

His body should have a deep but not broad chest, with ribs making a round rather than a flat side; his back should be short and level, and his loins strong and firm.

His tail should spring from the top, rather than the back of the body, it must not be thin or curved, but stout, and carried up straight.

The Rev. Dr Rosslyn Bruce, in his edition of *The Popular Fox Terrier*, quoted Edmund Burke, the great English orator of the eighteenth century, who wrote:

'Examine the head of a beautiful horse, find what proportion that has to his body, and to his limbs, and what relations these have to each other; and when you have settled these proportions as a standard of beauty, then take a dog, and examine how far the same proportions between his head and his neck, between those and his body, and so on, are found to hold; I think we may safely say, that they differ in every species, yet that there are individuals in a great many species, so differing, that have a very striking beauty; if this be allowed, it amounts to a concession that no certain measures' (measurements we call them) 'are necessary to beauty.'

How true this is. The great essential is to regard the dog as a whole, not in little parts. If the whole pleases, then minor faults must be forgiven.

CHAPTER 3

FEEDING AND GENERAL MANAGEMENT

I AM certain that the strength of a Kennel is in its bitches.

Beginners generally think that the best way of starting a Kennel is to buy an eight-week-old puppy and get it used to their ways, rear it well, win a few prizes with it and breed a champion from it! Sometimes they buy a dog puppy also, so that they can breed from the pair in the spring. I am afraid this does not usually work out very well, because no breeder of really good stock is at all likely to part with a puppy just out of the nest until he has made up his mind very clearly that it is not going to be one of his best ones, perhaps worth anything from £50 upwards in twelve months' time.

What, then, is the better way? One must realize that in entering a breed such as Fox Terriers, both Wire and Smooth, one is making a start in a lifelong and important pastime, with the prospect of both pleasure and possible profit in years to come.

Having decided that it is to be a Fox Terrier Kennel, it is advisable to attend as many shows as possible to help to form an opinion on the type of terrier preferred. The catalogues should be studied carefully to ascertain the names of the winning dogs and the Kennels which consistently produce winners and good, sound stock.

My advice is never to buy a dog or a dog puppy to start a Kennel. It is a big mistake to buy a dog with the idea of mating one's own bitch to him. It is far better to use the very best champion-bred dog in the country, one that is known to have sired winners, with the hope of breeding a dog puppy good enough to show and use at stud later on.

Presumably, all who are reading this book wish to breed Wire or Smooth Fox Terriers for show. If not for show, then good-looking, true-to-type, sound Fox Terriers. After a great many years of trying to do just this I am quite sure that the best way to start is to buy the best possible bitch that can be found for sale: that is, the best possible winning-producing blood-line. The bottom half of the pedigree is most important, i.e. the mother, grandmother and great-grandmother of the bitch intended to be bred from. My experience is that the bitch throws puppies about 75 per cent similar to herself. So breed from the best bitch that money can buy. If a fair price is paid for a good bitch it is a safe investment, whereas to buy a cheap brood bitch is just a waste of time and money: that is, if top-quality puppies are the object. On the other hand, if 'pot-boilers' are all that is required that is a different matter. By pot-boilers I mean puppies especially bred for the pet market. For these, a cheaper brood bitch, not up to show form, but strong and healthy, would suffice.

As an illustration of this principle I can recall the experience, many years ago, of a friend of mine. After several years of breeding Wires she found that she had not bred a single winner and could sell her puppies for only low prices as pets. She then came to me for advice. I inspected her bitches, considered their pedigrees and found them to be a very poor lot, which had originally been bought cheaply. My advice to her was to sell them all and put the money into one good, sound, well-bred bitch. This she did, and, again on my advice, mated the bitch to a fashionable dog of the day which had proved himself as a consistent sire of winners. The resulting litter included a dog puppy of considerable merit which had a lot of successes, culminating in a win at Cruft's Show, much to the owner's delight.

Had my friend continued until Domesday to breed from her former poor-quality bitches I am certain she would never have got out of the rut she was in.

Quite often people come to me to know how to start breeding good Wires, and I tell them: 'Buy the best bitch that you *cannot* afford. Put all your eggs in one basket.' I am sure that one good bitch is worth three average ones. So often I have heard it said, 'She is good enough to breed from.' I think this absolutely wrong. Always remember, 'The apple does not fall far from the tree.'

When choosing a brood bitch the first thing to look for, after the pedigree, is a bitch with as few faults as possible, structurally sound and of good character. These are essentials to me. I find light eyes and bad ear carriage two of the most difficult faults to breed out. Ears and eyes, if faulty, completely spoil the expression of a terrier, and I just cannot bear to see these two obvious faults every time I look at my dogs. Size is more easily rectified, and bone and substance can be helped by feeding.

Sometimes one finds a bitch that throws puppies like their sire, perhaps not the whole litter, but one or two. If anyone should be lucky enough to own such a bitch it is easy to produce winners in practically every litter simply by using the best dominant stud dog available.

If you do not wish to invest as much money as the likely cost of a good, young bitch the next best thing to do is to buy two bitch puppies from known, winning-producing stock. I say 'two puppies' simply because they are so much happier living together in a Kennel. If only one puppy is bought, and is kept in the house, it would have to spend a lot of time by itself, very unhappily, or take up a tremendous amount of one's time trying to keep it happy. Puppies are very demanding. When they reach the age of about six to eight months and, perhaps, start fighting each other, then the better puppy can be kept for breeding purposes. The not-so-good one of the two could be sold, possibly as a pet.

You might have to be content with an older bitch if you do not wish to cope with young puppies, or to spend the price of a young promising adult, but if you do this ask such leading questions as when was she in season last? Has she had a litter and proved an easy whelper and good mother? Has she been mated and missed? You might even find that she has had a litter of live puppies born, and that they faded out soon after birth. Any of these reasons might possibly be why she is for sale.

Another way to make a start is to acquire a bitch on 'breeding terms'. This means that the owner of the bitch lends her to breed a litter or several litters. The owner of the bitch usually requires the pick of the litter *at least*, and still retains the interest in the bitch for any future litters. It is advisable to have a written agreement on the form supplied by the Kennel Club for this purpose. Personally, I do not care for any such agreements. Pay and be paid is by far the best policy. Many so-called friendships have been broken through mixing business with pleasure. Suppose, for instance, that only one puppy is born; it may be thought hard if the owner of the bitch took the puppy, as he would be perfectly entitled to, according to the agreement. On the other hand, it is quite understandable that he should want it, because usually the only bitches allowed out on breeding terms are the ones the owner particularly wants stock from to retain the blood-lines in his Kennel. No! My advice is either to buy one or two puppies (as a gamble) or, as I have said before, and stress again, buy the best bitch that can be found at about ten or twelve months of age. Strong, healthy and handsome. She will cost more than an older bitch, but will be worth the extra cost.

Having decided, more or less, what is wanted, an appointment should be made with two or three Kennels to see what they have to offer. The great thing for an inexperienced beginner is to be frank, and to ask the breeder's advice. I am sure that all reputable breeders would do all in their power to advise a beginner on how to start this fascinating hobby, and also to help in any way possible after the purchase has been made. Be advised in the choice of stock by one of the older, established breeders, who will not only know the standard very thoroughly and how to apply it, but also advise about the different sires and dams whose names appear in the pedigree, for at least three generations back. This is most important.

I am afraid that many novices try to pit their wits against the old-established breeders. Nothing puts a breeder's back up more than the 'know-all' who goes to a Kennel and tells the owner, who has in all probability not only bred champions but also sold young stock as future champions, what is wrong with his stock and even how to improve it! Breeders of long-standing have forgotten more than most people will ever learn about dogs. I say 'of long-standing' purposely, because to continue breeding good dogs, or trying to, for a number of years one has to have patience, determination and a great love of animals. Keeping dogs is a full-time job, seven days a week, and is not all glamour and excitement. It makes me boil when someone comes up to me to say, perhaps when I have had a good win at a show, 'You *are* lucky!' Little do they realize the hours, and even days and weeks, I have spent getting my pride-and-joy to the peak of condition, both mentally and physically, before showing.

Having, after considerable care and numerous inspections, staked all available capital in the foundation stock, it is necessary to decide on the type of accommodation most suitable.

The best way of housing Fox Terriers naturally depends on the number of dogs to be accommodated. When only one or two dogs are to be kept, few kennels are needed.

Should you be lucky enough to have outbuildings of brick, which can be converted into kennels, you are more than fortunate. Windows, wooden floors and partitions can easily be inserted, and with a little imagination, and with the help of a jobbing builder, they can be converted into ideal kennels. All partitions used to separate kennels should be made of close-boarded wood, at least five and a half feet high. Terriers constantly try to get at each other, sometimes in a friendly manner, but more often to bicker and quarrel. It should be remembered that terriers are terriers, and like nothing more than to pick a quarrel, therefore every precaution must be taken to prevent their doing so. Wire mesh in between kennels or runs is not recommended, because should a dog put his nose or foot through the mesh while trying either to climb the wire, or to swear at his next-door neighbour, very serious bites can result. I have actually known this happen, a nose being practically torn off, and a front leg torn to ribbons.

Many types of kennels can be bought. Most are portable and consist of sections that easily bolt together. Some are designed to house several dogs and have a corridor giving access to individual kennels, rather like miniature loose-boxes for horses. I, myself, am not too fond of these, since when one wants to attend to a particular terrier it is impossible to do so without disturbing the remainder, with consequent noise, which neighbours may not appreciate. I like single buildings capable of taking one or two dogs living together, the other inmates of the Kennel are not, then, necessarily disturbed, especially if the buildings are arranged so that the dogs are out of sight of each other (see Fig. 7). All animals tend to be jealous, and there is bound to be trouble if for some particular reason attention has to be given to one of the dogs; the others are sure to resent it if the owner's visits are in full view.

Before deciding on the type of building to be purchased it is well to visit several manufacturers, either at their works or at the big shows where most firms have their products exhibited. Some have runs attached, but these are not essential, and expense can be saved by constructing the runs oneself.

The kennels should be constructed of wood, well built and strong, to keep out the wind and rain. Good windows are desirable, and strong, secure fastenings to all doors, both inside and out, are absolutely essential. The kennels should be raised off the ground on bricks, thus allowing air to circulate underneath and keep them dry.

The fronts of the kennels should have iron bars, and each kennel should be fitted with shutters and large windows. I prefer kennels to be high enough to allow me to stand up in them; this makes them much easier to keep clean. Each kennel has a wooden bench on which may be placed a tea-chest laid on its side and boarded across the front to prevent the bedding falling out.

If two terriers are living together it is as well to have two tea-chests for them to sleep in. There is always one master in any house, and

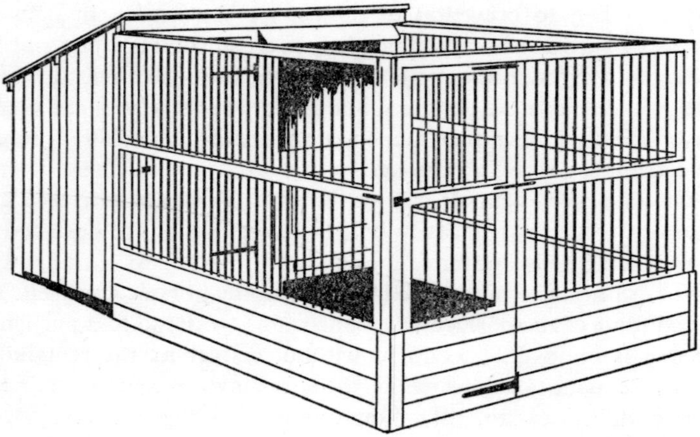

E. F. Hare & Sons Ltd.

Fig. 7 Single kennel and run.

quite often the 'king of the castle' will not let the other into bed. However, if two boxes are provided they nearly always settle their differences and curl up in one bed.

Besides the actual kennels, runs as large as possible are needed, enclosed by wire-mesh netting surrounds and access door. The wire netting should be of two-inch mesh and six feet high. The run should be laid with a slight fall to ensure drainage into a sump or drain. Concrete runs are preferred, because they can easily be brushed down with water and disinfectant every day, and so kept scrupulously clean. They are also good for terriers' feet, and help to keep the nails short.

The main objection to a cemented run is that it is liable to be cold and damp. This can be overcome by having, in the run, a wooden bench or a platform on which the terriers can sit in cold, damp weather when they are tired of playing.

A run laid around a well-grown tree gives endless amusement to the male members of the family! It also affords shade in the summer. Grass runs, I think, are undesirable, although many people prefer them, to save expense. I find that, sooner or later, a grass run becomes very foul and in wet weather it is useless for exercising purposes. Alternatively, in very dry weather the terriers have great fun digging holes, and the proud owner is continually worried by their dirty appearance. Some people use gravel for their runs, but here the difficulty is to prevent the small stones being eaten, and the soft surface spreads the feet. Also, they are very hard to keep clean.

A covered run is most desirable. One of my most prized possessions is a long building of brick construction with a felted roof, having full-length glass windows and a cemented floor. This covered run is quite the most useful of my buildings, because open runs are of little use during the severity of the winter.

Accommodation for puppies should be quite separate from that of older dogs, with wooden floors and plenty of room to play in. They, too, must have warm sleeping boxes and lots of bedding. The puppy kennels are much more comfortable with infra-red-ray lamps fitted in each, for warmth is most essential for the growth and development of all young stock. Of course, sunshine is best of all, but in the long winter days a lamp is the next best thing.

Another very useful building to own is a wooden or brick strcture, as large as possible, and high enough for anyone to be able to stand erect in. This is the stripping or grooming shed. The one I use for this purpose is 10 ft × 8 ft by 8 ft high at the ridge. It is made of close matchboarding and lined with hardboard. Down one of the long sides are placed 'show huts', four in all, raised on a platform thirty inches from the ground. The fronts of these have vertical iron bars about one and a half inches apart: if any wider the dogs are inclined to push their muzzles through the bars and so rub off their beards. For a dog being prepared for show this is a minor tragedy, because lack of furnishings on the face spoils the appearance of the head. Two good windows are set into the opposite side. Which reminds me, all my kennels have curtains to draw at night. These, I find, help the dogs to settle. Once the curtains have been put up, the dogs know it is time for bed, and there is not a sound from them until they hear someone moving about in the morning.

In this building there should also be a strong, steady table. Steady, because dogs hate being placed on anything that is not firm and gives them a feeling of insecurity. The important thing is to have the table made the correct working height. Electric light should be erected over this. It is surprising how many times it is needed, not only during the winter evenings but on dark days as well. Here, too, I have a small electric fire for use while I am working. I never leave a fire burning for any length of time when I am not there because I have a horror of wooden buildings catching alight; I have heard of dogs dying in fires, but never from cold (see Figs. 8 and 9).

A small cupboard or chest of drawers in which to keep brushes, tools and a few necessary bottles is most useful.

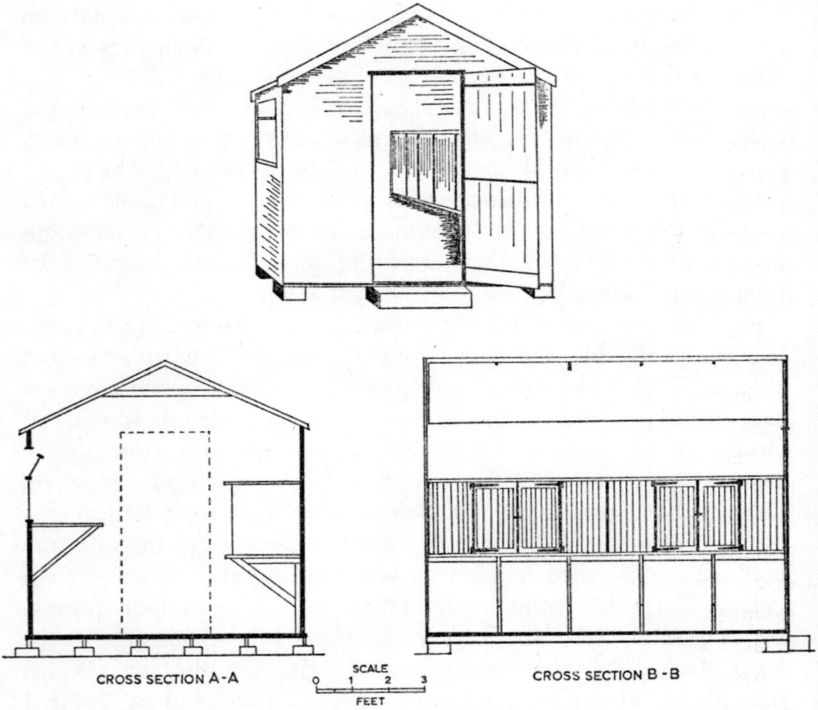

Fig. 8 Stripping and Grooming Shed

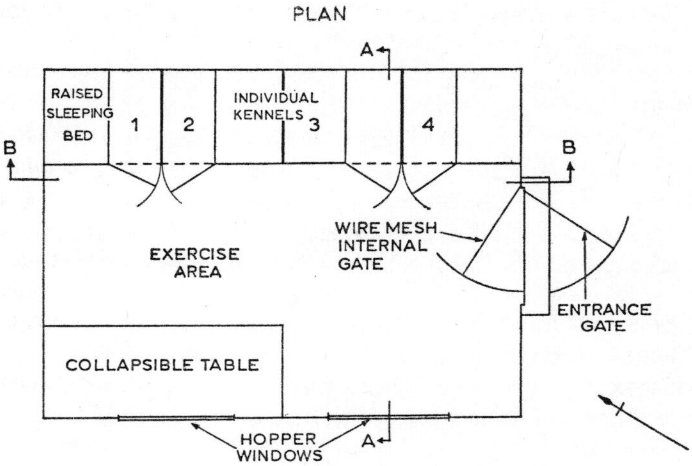

Fig. 9 Stripping and Grooming Shed

Two or three travelling boxes are necessary, and can easily be made by a handyman, or there are many different types on the market which can be bought quite cheaply. The best size for Fox Terriers, I find, is width 16 in., depth 22 in., height 19 in.—which is large enough for them to stand up and turn around in comfortably. The front should be half-boarded at the bottom, and the upper half should have vertical bars or strong wire. At the top of the boxes, down both sides, there should be ventilation holes.

The travelling boxes should be light and airy, with handles fitted for easy carrying. Dog boxes should be used as boxes, not as kennels. No animal should be left in a box all day. If these boxes are used for what they are made for they are ideal, as the dogs are much warmer in them when travelling, and seem to enjoy the comfort and security of a box of their own.

Newspaper should be used at the bottom of travelling boxes, with a blanket. Woodwool and sawdust is not practical, because when the dog comes out of the box, however careful one is, the woodwool and sawdust comes out too. Spare blankets should always be taken in case of accidents.

Another advantage of properly equipped travelling boxes is that when staying at an hotel the dogs will be little or no trouble to other guests, and most hotel managers usually accept well-controlled and well-behaved dogs in their boxes if they are to be kept in their owners' bedrooms.

Bedding

The tea-chests or boxes in all kennels should have fine woodwool as bedding. Having tried various kinds of bedding, I find woodwool quite the best. Straw so often contains thistles, and until it has been broken down it is harsh and spiky. Hay, in my opinion, is altogether too heating, and if soiled at all it steams and causes sweaty skins and feet.

Clean, white sawdust should be used generously on the floors of all kennels. Dark sawdust is undesirable for white dogs because the slightest moisture turns it black and quickly stains the legs or any part of the animal with which it comes in contact. The sawdust of any of the hardwoods, especially mahogany and oak, becomes very dark when damp. The best sawdust is derived from the soft timber of coniferous trees.

All kennels should be painted with bituminous paint or creosote on the outsides each year, not only to preserve the wood but also to make the whole place look well cared for and tidy. Nothing looks worse, to my mind, than buildings that have been neglected and allowed to become ramshackle old huts.

Kennels should be thoroughly cleaned at least once a day, beds shaken up, inspected, new woodwool provided when necessary and clean, dry sawdust spread over the floors. All excreta should be removed at once, as this is one of the main sources of infection. Food bowls must be washed after each meal.

Feeding and General Management

Here I am dealing with young adults from nine months onwards, as the rearing of puppies is described in another chapter.

First thing in the morning I open up all the kennels and put each dog out in either a yard or a covered run. If there are two living in one kennel, which is desirable, they naturally have their exercise together. Watch for any signs of diarrhoea or a loose motion, as this is an indication of the health of the animal. Diarrhoea is often due to unsuitable or too rich food.

Always feed each terrier separately, otherwise they will eat one against the other, and one of them, possibly, may eat more than is good for him. I cannot give exact quantities to be given at each meal because each dog varies in the amount it requires. Common sense must be used, but there is one golden rule which should be kept. That is, if any food is left in the dish after about ten minutes to a quarter of an hour it must be picked up and taken away. Food left over and allowed to stand in the kennel quickly makes a dog 'food-sick'.

For breakfast my young adults have corn-flakes and milk, the milk, preferably, previously boiled. They have it warm in winter and cold in summer. For any that need to put on weight I add a beaten-up egg to this meal on three days a week.

During the morning I bring each dog into my stripping shed, stand him on the table and 'go over him': that is, look at him well, handle and brush him. With practice it can soon be seen whether he is in good condition, with clear, bright eyes and tail wagging. I make a special point of noting whether his skin is itchy, and if it is I apply a soothing lotion at once. The great thing is to nip any little upset in the bud. My husband laughs at me when I say, 'So and So looks pale this morning'! It invariably is so, and usually the dog in question shows the real symptoms a little later on—if a bitch, she, perhaps, may just be coming in season or feeling a bit off colour.

Each day the feet should be studied, especially in between the toes. Any dirt left in between the pads quickly makes the foot sore and the dog lame.

After I have 'gone over' each dog carefully, and attended to him in any way that is necessary, I put him in one of the show boxes in the stripping shed, previously described (Fig. 8), to sit on a piece of blanket to watch me strip or trim one of the dogs being groomed or prepared for show. He enjoys the company while I am working.

After another run, just before lunchtime, they are all put into their respective kennels to sleep. This is important both for the dogs and their attendant, who, after all, must have a bit of peace and quiet. Also, it is no good feeding a dog well if he does not have sufficient sleep during the day in comfortable quarters. A dog continually on the run will never put on sufficient condition for show. I find that terriers that are well fed, exercised sufficiently and given cosy beds are happy, contented and sleep a good deal.

In the evening they all have another exercise in the runs, and any that are being trained for show are exercised on a lead up and down the lawn in an enclosed space. Sometimes they are brought into the house to get used to all kinds of noises and placed on a table time and time again to teach them to stand correctly while being handled. After this they are all put into their kennels where they have their main meal of the day. This consists of brown bread which has previously been soaked in gravy, biscuit meal also previously soaked and plenty of cooked minced meat. Occasionally I feed raw meat to in-whelp bitches, as some seem to crave for it, but I do not feed it as a general rule. I find it often causes diarrhoea, and seems to be indigestible. I know it is said that 'raw meat is their natural food', but I contend that we do not keep our dogs in a natural way in these days. If dogs were living naturally, hunting for their food, they would take an enormous amount of exercise both before and after eating their kill. They would not eat just the lean meat either, they would eat fur and all! Fresh water should always be kept available, and in the summer this must be changed frequently and never placed in the sun.

One cannot, of course, keep to a strict routine every day, but it is wise to keep to one as nearly as possible. Such things as matings and showing people the stock have to be fitted in when convenient. When I put a dog back into a kennel I always give him a titbit, such as a sweet biscuit or several Vetzymes.

Remember that these are kennel dogs, not pets. As such, they should live in kennels and should not be brought into the house for any length of time, otherwise they will not settle contentedly in their kennels. This is sound advice, which I am afraid I frequently break for old favourites.

CHAPTER 4

WIRES: DOMINANT SIRES AND DAMS

BEFORE the First World War the most consistently successful Wire Fox Terrier stud dogs were the 'Of Notts' owned by Her Grace the Duchess of Newcastle. Ch. Cackler of Notts was the sire and grand-sire of an astonishing number of great terriers, such as Ch. Briar Cackler, Ch. Commodore of Notts, Ch. Captain of Notts, Ch. Raby Coastguard and Ch. Dusky Cracker. It is quite safe to assert that there is not a pure-bred Wire Fox Terrier in the land which does not descend from the 'Of Notts' terriers.

Mention must also be made of Mr S. G. Fildes's famous 'Wyche' Wires, handled by the professional handler, the late Arthur Cartledge, whom I knew very well. Two of these terriers in particular had wonderful show careers. Ch. Wyche Workman, born July 1917, won, in 1920, six successive Challenge Certificates in four months and Ch. Wyche Warm, born April 1923, won his first four Certificates in six weeks. Ch. Wyche Workman sired Ch. Wyche Wrangler, Ch. Wyche Witchery and Ch. Wyche Wondrous. Ch. Wyche Witchery produced Ch. Wyche Warm.

Ch. Fountain Crusader was one of the very first champion dogs I saw winning in London. I was young at the time and he impressed me very much. On looking back over the years I often wonder how these famous terriers of the past would compare with our present-day champions.

In the 1928 Wire Fox Terrier Association's Year Book Her Grace the Duchess of Newcastle wrote:

'Coats, feet and hindquarters all want great attention. We have evolved a very high class quality animal in the present-day Wire, but I am beginning to ask myself, are we getting too far away from the old standard? "Bone and strength in a small compass", and "A terrier should stand like a cleverly made hunter".

'Well, I believe we are getting terriers too much like a great slashing blood horse of the Aintree type, and I'm sure with too much size.'

In my opinion these remarks apply equally today.

Ch. Fountain Crusader sired several champions and will always go down in history as the father of Ch. Talavera Simon, who proved himself to be the greatest sire of all time. Simon sired eleven champion sons and twelve champion daughters. He was a terrier of the correct size, richly hound-marked and a great personality.

One of the greatest judges of dogs, the late Holland Buckley, reported on Simon as follows: 'Talavera Simon is a very beautiful Terrier, indeed, as sound as they are made, with the most perfect body, coat, bone and head imaginable. He is as near faultless as they come. If he does not turn out one of the greatest Sires, not only myself, but others fit to form an opinion, will be truly amazed.' How true this prediction proved to be.

At the age of eleven years Simon sired Ch. Darroch May Queen, she being the last of the champions he sired. 'The Queen' had all-tan markings, which strongly appealed to me. Many people much prefer hound markings, and very few ginger-marked Wires are seen today.

Miss J. Abel, writing in the Wire Fox Terrier Association's Year Book, in 1945, reported on her thus: 'Ch. Darroch May Queen—She is grandly made, excelling in neck and shoulders and shortness of back, and is the soundest of movers.'

I, myself, owned several daughters of Simon in the early 1930's, all very similar in type, hard-bitten terriers with excellent coats, and very sound.

The female Simon blood-line is exceptionally strong, since from his daughters came Int. Ch. Gallant Fox of Wildoaks, Ch. Beau Brummel, Ch. Talavera Jupiter, Ch. Talavera Nigel and Ch. Crackley Startler, whose names appear in most present-day pedigrees.

Ch. Crackley Startler I particularly remember, having seen him shown by the greatest professional handler the Wire ring has ever known, Mr J. R. Barlow. Today Bob Barlow is known as Mr Fox Terrier! Ch. Crackley Startler had an even-marked tan head, nearly all-white body, with a black patch on his loin and one at the root of his tail. I remember that one of his ears was half marked tan and white, which is somewhat unusual with an even-marked head; usually, the ears are all tan. He was exported to America at an early age, which I felt was a pity, because during the short time he was in this country he sired four champion sons and four champion daughters, with many winners besides.

I believe I am correct in saying that Bob Barlow took Startler over to America and returned to England bringing with him Ch. Gallant Fox of Wildoaks, owned by Mrs Bondy of America. He also brought with him Ch. Beau Brummel of Wildoaks and Ch. Bobby Burns of Wildoaks, who quickly became champions.

Ch. Gallant Fox, Ch. Crackley Startler, Ch. Talavera Jupiter, Ch. Beau Brummel and Ch. Talavera Nigel, handled to their championships by Mr Barlow, all proved to be dominant sires. Of these, Jupiter sired sixteen champions, Gallant Fox fifteen and Beau Brummel thirteen champions.

Ch. Talavera Romulus, son of Ch. Talavera Simon, has played a big part in producing the champions I myself have bred. Penda Pompilius was actually sired by Ch. Talavera Romulus. Pompilius was a biggish, nearly all-white dog, with a little tan about his head He had the most beautiful neck and shoulders, an exceptionally hard coat, and was very sound and a grand mover. He was a 'war baby', which meant he could not be shown at championship shows, as none were held. I do not think he would have become a champion; he was, in my opinion, too big. He won a great many awards at the smaller shows, including several Best-in-Shows. He sired two champion bitches, from whom sprang Ch. Wybury Penda Quicksilver, the dam of several champions to which I will refer later.

The first dog to be made a champion when the big shows started again after the last war was Ch. Crackley Straightaway. I remember him as a correct-sized dog, with a very long head and great substance. Colonel Phipps of the famous 'Talavera' Fox Terriers judged him at the Wire Fox Terrier Association's Show in 1946 and reported in the Wire Fox Terrier Association's Year Book on him as follows:

'Crackley Straightaway is, I think, one of the best terriers I have ever seen. I had heard lots of criticisms of him, but I failed to find one that I felt was in the least bit justified. I don't want to alter him, anyway. I suppose great dogs, like great men, always attract the "crabbers" and fault-seekers. He has stud dog written all over him. A very, very, great Terrier in my opinion.'

Ch. Crackley Straightaway sired Ch. Polo Fireaway, and among the many winners he also sired was Newmaidley Ceasar, who has had great influence in the 'Newmaidley' Wires owned by Miss L. Beak. Linda Beak has owned and bred several champions, namely Ch. Newmaidley Hob, Ch. N. Cleopatra, Ch. N. Dancer and Ch. N. Treasure. Ch. Polo Fireaway, owned by Mr Charles Cole, sired Ch. Arley Miss Quality and Ch. Graig-y-Dan Fearnought.

In 1947 five more dogs won the coveted title of champion. They were Ch. Clarington Contender, Ch. Holmwire Hyperian, Ch. Crackley Sailaway, Ch. Travella Strike and Ch. Weycroft Woolcomber. These were all attractively hound-marked dogs with the exception of Ch. Weycroft Woolcomber, who was lightly hound-marked. All were very similar in type and size, with the exception of Ch. Crackley Sailaway, who was a much smaller dog.

Here I would like to relate a lesson I learnt from Mr Barlow, not so much about the correct size for Wire Fox Terriers, but about manners! Mr Barlow, indisputably the most knowledgeable terrier man of modern times, was then handling his own Ch. Crackley Sailaway, and also Eden Autocrat, professionally, for his client, Mr F. Robson. I, insolently (so Mr Barlow now tells me) asked, 'Which, Mr Barlow, is the correct size for a Wire Fox Terrier, Eden Autocrat or Crackley Sailaway?' Eden Autocrat was, to me, a really big dog and Sailaway was on the small side for some. As quick as a flash came the reply, 'Whichever one you own, madam!' Yes, I was a 'madam' then! I have been brought up with the idea that Wires *must* be the correct size; I do not like dogs that are too big. It is interesting to note that although Eden Autocrat remained just a big winner, Sailaway quickly became a champion. Eden Autocrat sired Ch. Waterton Wendy and Ch. Wycollar Duchess, the latter of unusual markings, a black marked head and body. Ch. Crackley Sailaway was the sire of Castlecroft Cleanaway.

Ch. Travella Strike started his short, sharp, successful show career in 1947. He was born 26 September 1946. By Travella Sensation ex Travella Gloria, descending from Ch. Gallant Fox and going back to Ch. Travella Simon through the bitch line. He caused quite a stir in Wire Fox Terrier circles when he came out. He won his first Challenge Certificate, under Leo Wilson, when only ten months old, his second two months later, when the judge was George Wallwork, and the third and qualifying Challenge Certificate under Mrs E. M. Lester on 27 September 1947, thus becoming a full champion one day after his first birthday. In the report of him after the show the judge, Leo Wilson, wrote of Travella Strike: 'His beautiful quality head, and more especially his lovely neck and shoulders, were an object lesson, and the line from the back of his head to his tailroot was particularly attractive.'

There is no doubt that Strike has been as equally sensational as a sire as he was a show dog, stamping his beautiful head and expression, plus the all-important quality, on all his stock. The 'Travellas' owned by W. Browne-Cole certainly had a phenomenal run of successes in the early 1950's and it is safe to say that the descendants of Strike have always to be reckoned with whenever they are shown at the present time.

Ch. Travella Strike sired the following champions: Ch. Travella Skyliner, Ch. Meyrex Travella Reliable, Ch. Travella Sizzler, Ch. Travella Skyflyer, Ch. Arley Adorable, Ch. Roundway Cats Cradle, Ch. Travella Wildcroft Superb, Ch. Travella Starshine, Ch. Masterlea Lustre, Ch. Axholme Double Strike, Ch. Roundway Strike-a-Light, Ch. Rawdon Rhoda, Ch. Striking of Laracor, and Ch. Harrowhill Strike Again, and innumerable well-known winners.

Mr W. Browne-Cole also owned the following champions: Ch. Travella Quick Decision, Ch. Travella Superman, Ch. Travella Allure, Ch. Travella Suredo, Ch. Travella Sureset, Ch. Travella Silk, Ch. Travella Supercatch and Ch. Travella Superstar. Of these, Ch. Travella Starshine sired Ch. Travella Suredo and Ch. Travella Silk. Ch. Travella Sureset was the sire of Ch. Travella Supercatch. Ch. Travella Skyflyer was the sire of Ch. Travella Superman and Ch. Travella Allure. Ch. Travella Wildcroft Superb was the sire of Ch. Sunnybrook Superjet and Sunnybrook Special Choice. Ch. Striking of Laracor was the sire of Ch. Whitwyre Field Marshal, who has sired Ch. Florate Fondah and Ch. Whitwyre Even Money. Ch. Axholme Double Strike was the sire of Ch. Shoemans Welland Winstan, Ch. Lyngarth Social Call, Ch. Crackyn Corrector and Ch. Penda Purbeck Repeat.

It is therefore obvious that a number of successful Kennels of today owe a lot to Ch. Travella Strike. Many of the winners produced at the Roundway Kennels, owned by Mrs J. Creasy, have Strike as their sire or grandsire. These include Ch. Roundway Cats Cradle, and Ch. Roundway Strike-a-Light, who sired Ch. Roundway Bellbhoy and Ch. Lemonford Lollipop of Roundway. Ch. Roundway

Bellbhoy sired Ch. Clennon Chime, winner of nine Challenge Certificates, and many more winners.

When one thinks of such good dogs as Ch. Weltona Revelation, Ch. Weltona Flame, Ch. Weltona Exemplar, Ch. Weltona Exelwyre Dustynight, Ch. Weltona Realstance, Ch. Weltona Exempless, Ch. Weltona What's This, Ch. Weltona Miss Sundance and Ch. Weltona Platta Princess, one realizes that their owner, Mr A. Churchill, must be mentioned as a stalwart of the breed. Ch. Weltona Revelation, one of the first dogs to become a champion after the last war, sired three champion dogs and one champion bitch, and is the grandsire of many champions.

Ch. Weltona Exelywre Dustynight deserves particular mention as the sire of the superlative Ch. Madam Moonraker, who before leaving for America won several Best-in-Show awards at our biggest championship shows. I remember awarding her the Championship at Bath, and how she completely pleased me. Later in the day she won Best-in-Show all Breeds, under the Earl of Northesk. Dustynight was the sire of my home-bred bitch Ch. Penda Peach, who won Best-in-Show and the Gold Cup at the Fox Terrier Club Show in 1956. Dustynight has certainly had a great influence on the breed.

In 1945 Mrs Dorothy White bred Ch. Wyretex Wyns Tuscan, whose sire was Culverbrook Tuscan (Ch. Talavera Romulus) ex Wyretex Wyns Thralia. Tuscan had a beautiful long, lean head, grand legs and feet, and was not a big dog. He proved a very good sire indeed. His English champion sons and daughters are as follows:

Ch. Weltona What's This, Ch. Wyretex Wyns Wundar, Ch. Penda Callern Melody and many Continental and American champions. Melody had rich, all-tan markings, and is the only 'ginger' bitch to become a champion since Ch. Darrach May Queen nearly twenty years ago. Ch. Wyretex Wyns Wundar (sire Ch. Wyretex Wyns Tuscan ex Ch. Wybury Penda Quicksilver) I bred on 4th May 1955, and sold to Mrs White as an embryo champion when he was about five months of age. He was a biggish, nearly all-white dog, excelling in head, neck, shoulders, coat and movement, with a distinct resemblance to his forbear Penda Pompilius. Wundar has proved himself to be a great sire, his English champions being:

Ch. Zeloy Endeavour, Ch. Cudhill Kalipso, Ch. Steetonian Skipper,

Ch. Helenstowe Pied Piper, Ch. Wyretex Wyns Wunderful and far too many overseas champions to name. It will be seen from this list that he is the grandsire of several champions, among them Ch. Zeloy Emperor and Ch. Rogerholme Radiance.

In 1950 Mr C. H. Burton bred Ch. Burtona Betoken, and later sold him to Mr J. H. Pardoe owner of the famous Cawthorne prefix. Betoken proved a great stud dog and was an attractively hound-marked dog of the correct size. His winning progeny include: Ch. Cawthorne Climax, Ch. Cawthorne Chloe, Ch. Kenelm Supremacy and Ch. Steelholme Sheena. Ch. Cawthorne Climax, a practically all-white dog, is the father of Ch. Emprise Sensational, Ch. Climax Token, Ch. Windlehurst Susan and Ch. Penda Cawthorne Cobnut. Ch. Kenelm Supremacy's untimely death when he was still a young dog robbed us of a good stud dog. During the comparatively short time he was at stud he sired Ch. Penda Oregon Witchcraft, Ch. Kenelm Miss Supremacy, Ch. Oregon Highspot and Ch. Taywell Tearaway.

I purchased Penda Cawthorne Cobnut after judging him at the Fox Terrier Club Show in 1958. He did not win there, being placed second in his class, because he had not bodied up, but I liked his size, bone and shortness of back. My prefix was added to that of Mr Pardoe's and he became Ch. Penda Cawthorne Cobnut. Cobnut has sired Ch. Baros Jewel, Irish Ch. Baros Cobbler, Ch. Wyrecroft War Bonus and Ch. Penda Peerless. Ch. Penda Peerless is proving himself to be an outstanding stud dog, having already sired Ch. Penda Daleskirk Caress, winner of Supreme Best-in-Show at Bath and the Gold Cup at the Fox Terrier Club Show, 1962, and the Championship at Cruft's, 1962 and 1963 shows. Also Ch. Penda Tavatina, winner of the Gold Cup 1964.

Last, and by no means least, we come to Ch. Crackley Standard, who has sired Ch. Crackyn Correct, Ch. Crackyn Captivator, Ch. Crackyn Caprice and Ch. Wicklewood Candybar.

Ch. Crackley Standard, who became a champion in 1956, was again a correct-sized terrier, most attractively hound-marked, and a dog that will surely be remembered as being the sire of Ch. Crackyn Cockspur. Cockspur was handled by Mr Barlow to win no fewer than twenty-two Challenge Certificates, Best-in-Show at Cardiff Championship Show, Best-in-Show at Birmingham Championship Show and Supreme Champion at Cruft's, 1962. He was then sold to M. Lejeune of Belgium, where he won Best Terrier at the International Championship Show in France, Best-of-all-Breeds at the International

Championship Shows in Belgium, Holland, Italy, Germany and Luxembourg; also, Best-in-Show-all-Breeds at the Swedish Kennel Club Show in Stockholm.

Cockspur is a correct-sized terrier, most attractively hound marked, a very long head, short back, and the most beautiful legs and feet possible, with a good wire coat right down to his feet—a rarity these days. He had to be handled to be appreciated; in fact, an outstanding dog, very hard to beat.

England's loss is Belgium's gain, as has been so often the case in the history of our breed, large numbers of our top champions having been exported to all parts of the world.

It is difficult to say which bitches have influenced the breed most over the years. It must be remembered that breeders, having campaigned their bitches to championship status, which costs a great deal of money, frequently export them instead of breeding from them. Consequently, very few champion bitches are bred from, and I feel sure this is the reason for so many people believing, and saying, that champion bitches do not make the best broods.

Another point which has occurred to me is that top show bitches are seldom mated much before they are two and a half to three years of age because showing cannot be combined with breeding. The show bitch's breeding period is therefore considerably shortened and tends to make people believe that the best terriers are bred from non-show specimens.

International Champion Gains Great Surprise in 1934 was described as one of the greatest show bitches of all time, and if not the first, was one of the first champion bitches sired by Ch. Talavera Simon. She, in turn, was the mother of International Champion Gallant Fox of Wild Oaks and Ch. Beau Brummel of Wild Oaks, two very famous sons, and she therefore must go down in history as a truly wonderful bitch.

Kersley Pandy was another famous brood bitch. I do not think she was considered a good show specimen, but she produced many champions.

Talavera Patchworth, mated to Ch. Talavera Simon, produced Ch. Epping Editor.

Another lovely bitch, Ch. Castlecroft Content, sire Ch. Beau Brummel, mated to Ch. Gallant Fox of Wild Oaks, produced Ch. Castlecroft Contender.

In 1940 a Gallant Fox bitch, who was mated to Ch. Crackley Supreme Again, bred Ch. Crackley Striking.

Ch. Kirkmoor Carefree, owned by W. Mitchell, was the first bitch to be made a champion after the last war.

At the end of this book the pedigrees of all the champions that have been made since 1946 to 1964 are listed, and I do not therefore propose to dwell here on all the good bitches that have done well at shows and also produced champions.

Just a few, however, I feel I must mention as being outstanding specimens of their day.

Ch. Stocksmoor Sharecroft Select, bought by Mrs Creasy, I remember as being a truly lovely bitch—she became the Dam of Ch. Roundway Cats' Cradle.

J. Hamilton bred the attractive Ch. Burntedge Besum. Besum made a name for herself by being the mother of Ch. Burntedge Ballet Girl.

It may be of interest if I relate an incident that happened when W. Mitchell, one of our top professional handlers, was showing Burntedge Ballet Girl and I was showing my Wybury Penda Quicksilver at the Wire Fox Terrier Association's Show in 1953. Mr 'Towyn' Edwards was the judge, the great all-round terrier expert. To win at this show was, and still is, considered a prize of much importance.

The open bitch class had about fourteen entries and 'A.J.', as Mr A. J. Edwards was always known by his friends, took quite a long time to place his 1st, 2nd and 3rd, there being some very good bitches present. After placing Burntedge Ballet Girl 1st and Quicksilver 2nd, Mr Edwards stopped looking at the dogs, stood back and remarked to the crowded ringside that he had left several well-known winners 'out of the money' because he considered Burntedge Ballet Girl and Quicksilver to be of the correct size, the others being of a bigger, more racy, type. Both Ballet Girl and Quicksilver were thought to be too small by quite a number of our judges. They certainly were not weeds, and had plenty of substance. Both became champions and both were the dams of champions. Ch. Burntedge Ballet Girl, mated to Wyrecliffe Sunnybrook Spitfire, bred Ch. Kirkmoor Crocus. She was exported by W. Mitchell to America at a remarkably high figure, we are told.

Ch. Wybury Penda Quicksilver certainly must be mentioned as one of our best winning-producing bitches.

She had quite an interesting history, and I would like to explain why she carries the late Mr Berry's prefix as well as my own, although I bred her.

When Quicksilver was about ten months old I entered her in the puppy bitch class at Leeds Championship Show. This meant my travelling all night by train with her, arriving at Leeds at about four o'clock in the morning, to a cold, empty station. No restaurants were open and no porters were available to carry the boxes. However, we found a waiting-room open, and there we parked the dogs in their boxes while a few fellow travellers and myself wandered about Leeds until we found an all-night café, where we had hot drinks and sandwiches.

At about nine o'clock we set off to the show. On looking at the puppy bitch class entry I saw that only four were entered. As Quicksilver had already won three first prizes in very strong competition at the Fox Terrier Club Show, I was most hopeful. She was in tip-top condition, with a really good coat. Several onlookers admired her greatly while I was making final preparations, and into the ring I went —hope high indeed!—she came last in the class! As I came out of the ring, tired, cold and my hopes shattered, Mr W. Berry, who owned the famous prefix 'Wybury', asked me if I would sell her to him. This I did in a moment of despondency. I forgot the truth of the adage:

Fair lilies bloom 'mid dank and marshy stenches
Sweet deeds are done by most ill-favoured wenches
And cups reward some worst dogs on the benches
So all may hope!

Quicksilver was shown for Mr Berry quite a bit, but did not do very well for him. She was a possessive bitch and, as I had nursed her through hardpad, probably required more fussing than he was able to give her, and which she had with me.

When she was two years old Mr Berry telephoned me saying that he did not wish to breed Wires but would prefer to concentrate on his well-known Scotties and would I like to buy her back at a nominal sum. I was delighted to do this and had great enjoyment in making her into Ch. Wybury Penda Quicksilver.

She proved a good brood bitch. Her champion sons and daughters and grandchildren will certainly appear in pedigrees for many years to come.

CH. PENDA PEACH	Ch. Weltona Exelwyre Dustynight	Middleforth Tuscan	Culverbrook Tuscan
			New Lane Peggy
		Juliet of Exelwyre	Exelwyre Diplomat
			Exelwyre Jewel
	Ch. Wybury Penda Quicksilver	Ch. Penda Blackwell Revelation	Ch. Weltona Revelation
			Edenholme Elfreida
		Ch. Penda Hieover Warrior	Penda Pompilius
			Hieover Music

For her first litter she was mated to Ch. Weltona Exelwyne Dustynight, to whom she had five puppies. The dog puppies all seemed too big for me, consequently I kept only one bitch from this litter. She became Ch. Penda Peach, who won the Gold Cup for Best-in-Show at the Fox Terrier Club Show in 1956. She also made history by being the first Wire Fox Terrier to be the third champion bitch in direct line of breeding.

For Quicksilver's second litter I mated her to Mrs D. White's Ch. Wyretex Wynns Tuscan, to whom she had only two dog puppies. One was sold as a pet to France at an early age, the other was bought by Mrs White at about five months of age and quickly became Ch. Wyretex Wyns Wundar.

Next I mated Quicksilver to Ch. Crackley Security. She had five puppies, but only one bitch survived, which later became American Ch. Penda Peachblossom. She was not shown in England, being sold at about ten months of age. A year later 'Silver' was again mated to Ch. Crackley Security and brought up five more puppies. One dog, again large enough, called Penda Silver Prince, was exported to South Africa, where he became a champion and won Best-in-Show, all breeds, at their championship show. Penda Silver Lining, also from this litter, a small, heavily hound-marked dog, was also exported, this time to the U.S.A. Penda Silver Lady, the litter sister, I kept and showed here, winning the Challenge Certificate and Best-of-Breed and one Reserve Challenge Certificate before selling her. She also won a Reserve Challenge Certificate for her new owner before being retired for maternal duties. She, too, has bred winners.

Quicksilver was next mated to Ch. Axholme Double Strike. This resulted in two dogs and two bitches. Both dogs were too big for me; one I sold to Ireland, and he, in turn, sired the lovely bitch Ch. Baros Romance and several good winners. He was a hound-marked dog. The other dog, called Penda Pomp (he greatly resembled Penda Pompilius, being nearly all white), I sold. He did not please, and came to an untimely end—but not before he had sired winners.

By this time I realized that Quicksilver would not breed a correct sized dog unless mated to a real cobby huntsman type of terrier. Having bought Cawthorne Cobnut (who later became Ch. Penda Cawthorne Cobnut) she was mated to him. As Quicksilver was now just under eight years of age, this was to be her last litter. She whelped two dogs and two bitches. One small, nearly all-white bitch

PEDIGREE OF CHAMPION PENDA PEERLESS

Sire CH. PENDA CAWTHORNE COBNUT 1062 A.R.	Ch. Cawthorne Climax	Ch. Burtona Betoken	Burtona Bo'sun
		Cawthorne Twynstar Actionette	Torkard Countess
	Cawthorne Ready Maid	Florate Frontpiece	Ch. Twynstar Accurist
		Warm Welcome	Castlecroft Cleanaway
Dam CH. WYBURY PENDA QUICKSILVER 1665 A.L. (Dam of five champions)	Ch. Penda Blackwell Revelation	Ch. Weltona Revelation	Culverbrook Tuscan
		Edenholme Elfreida	Hoddlesden Lady
	Ch. Penda Hieover Warrior	Penda Pompilius	Ch. Talavera Romulus
		Hieover Music	Wyretex Wynsdainty

I sold when young. The other bitch I still have. She is attractively hound-marked, and is called Penda Painted Lady. In her first litter she bred Penda Perfect Lady, a championship show winner, before being exported.

Of Quicksilver's two dog puppies, the all-white one became Ch. Penda Peerless, mentioned above, who won Best-of-Breed at Cruft's Show in 1961. I consider Peerless to be the best Wire I ever owned and his pedigree may be of interest. His litter brother, Penda Caesar, is still with me and is proving a prolific sire. He is a winner at championship shows and a great personality.

So you see, Ch. Penda Quicksilver proved herself to be a bitch in a million, one that has not only produced champions herself, but whose children's children are doing so today. I cannot recall any other champion Wire bitch who has been the dam of three English champions and several foreign champions.

Finally I must mention the record-breaking Ch. Gosmore Kirkmoor Tessa. Mrs A. B. Dallison bought this bitch from W. Mitchell when still a puppy, I believe. She was bred by Mr W. Ratcliffe, sire Exelwyre Mooroak Aristocrat, dam Brigston Carosel Miss Fonda.

Gosmore Kirkmoor Tessa made her début at Manchester Championship Show in 1963. I am told she caused quite a stir there (alas, I was in hospital at the time), as not only did she win the Challenge Certificate from the novice class, but she also won Best-of-Breed, and the coveted Best-in-Show award. During 1963 Ch. Gosmore Kirkmoor Tessa won 15 Challenge Certificates, which is considered to be a record in the breed. It gave me great pleasure to award her the Challenge Certificate at Cruft's 1964, and to see her go on to win the terrier group. At the time of writing she has won 19 Challenge Certificates, 3 Best-in-Shows and 3 Reserve Best-in-Shows.

Her picture, which shows her excellent type and very attractive markings, is reproduced.

As well as Ch. Gosmore Kirkmoor Tessa, Elexwyre Mooroak Aristocrat has sired Miss B. Cliff's Ch. Wyrecliffe Satellite of Senegal and Ch. Gosmore Elexwyne Diamond.

Mr E. Robinson's Ch. Zeloy Endeavour has already sired Ch. Zeloy Crusador, Ch. Zeloy Emperor and Ch. Gosmore Mariebel Tina.

During the autumn of 1963 Penda Tavatina made her appearance and has since proved to be another 'star'. Bred by Mr R. Davison, sire Ch. Penda Peerless, dam Tavaprim.

Beautifully presented and handled by W. Mitchell, she won her first Challenge Certificate at Manchester 1964, her second at the Northumberland and Durham Fox Terrier Club Show, where she was also Best-in-Show, and she gained her third Challenge Certificate and her title at Bath Show in 1964. At the National Terrier Show she was undefeated, again finishing Best-of-Breed.

At the 1964 Championship Show of the Fox Terrier Club for both Wire and Smooth Fox Terriers, Champion Penda Tavatina finished her show career in England, winning the Challenge Certificate, the Club's Gold Cup for Best-of-Breed, and the award for Best-in-Show.

Shortly afterwards she was flown to the United States of America. Tavatina certainly had a brilliant, though short, show career in England, and it is to be hoped that she will do equally as well for her new owners.

It is very evident that Ch. Zeloy Emperor will have a lasting influence on the breed. During the last four years he has sired no fewer than nineteen English champions: ten dogs and nine bitches. This is truly a great record of which Mr E. Robinson—'Robby' to his friends—must be justly proud. Recently Mr Robinson had the honour of being President of the Wire Fox Terrier Association.

Ch. Penda Peerless when mated to Penda Purbeck Deborah produced Penda Popular (later re-named Wyrecroft Penda Popular) and also U.S.A. Ch. Penda Peerage. Popular won one Challenge Certificate before being exported to America. Before he left England he sired the two champion bitches Ch. Penda Peppermint and Jokyl Wyrecroft Jemini from Ch. Wicklewood Candybar. Ch. Penda Peppermint remains in my kennel to carry on the Quicksilver line.

It is interesting to note that, besides Ch. Wicklewood Candybar, Ch. Mitre Miss Strike, Ch. Weltona Platta Princess and Ch. Worsbro Oladar Royal Maid have each bred champion sons and daughters.

Ch. Worsbro Oladar Royal Maid when mated to Worsbro Betoken Again produced the quite outstanding Ch. Penda Worsbro Whistler and Ch. Penda Worsbro Weasel in one litter. Mr F. Robinson must be congratulated on breeding two such beautiful Terriers.

Ch. Penda Worsbro Whistler won the Fox Terrier Club's Gold Cup for Best-in-Show in 1968 and together with Ch. Penda Peppermint won the American Cup for best brace in Show. Ch. Penda Peppermint also won Best Bitch in the group at the Birmingham Dog Show Society's Championship Show, 1968.

CHAPTER 5

SMOOTHS: DOMINANT SIRES AND DAMS

A MAN without a hobby is like a dog without a tail. Whether the hobby is collecting old china or postage stamps, the quest will enrich the collector's experience and give him an added interest in life. But if one is to collect something why not let that something be an animal with life and movement? A well-bred hunter is one of the most shapely animals we possess, and a Fox Terrier is moulded on similar lines. Why not, therefore, choose as a hobby the breeding and exhibiting of Fox Terriers?

The Fox Terrier knows how to behave in any society. If well trained he is not a yelper but barks only when necessary. He loves motoring, and will sleep in his own box in strange hotels without a murmur and will come out in the morning ready for anything. He is a gay, happy dog that one can be proud to own.

There is no doubt that correct terrier temperament in the breeding and showing of Fox Terriers is equally as essential as type and quality. One can take a lovely-looking Smooth—or Wire for that matter—into the show ring, but without correct terrier character one most certainly will suffer continual disappointment. A gay, happy terrier is a joy to behold, and affords endless pleasure. Without a proper temperament a terrier can be an abject creature, either shy and timid, with its tail between its legs, or a snapping, snarling little beast which makes itself an utter nuisance.

If we allow our minds to dwell on the outstanding Fox Terriers of the past we must come to the inevitable conclusion that in every case we remember them by their great presence, by their great vitality and dominance in the show ring, and by their ability in transmitting these qualities to their offspring.

Mr Wardle's famous painting of the late Francis Redmond's 'The Totteridge Eleven' depicts a team of most beautiful Smooth Fox

Terriers, all well-known winners, including several champions. A wonderful group, which shows the quality, charm and character of the Smooth variety of Fox Terrier. This beautiful picture is now the property of the Kennel Club, with whose permission it is reproduced. The late Francis Redmond bred ten champions. A great achievement.

Her Grace the late Kathleen Duchess of Newcastle, who was President of the Fox Terrier Club for many years, bred nine champion 'Of Notts' Smooth Fox Terriers, and owned many more. These include such famous names as Ch. Chosen Don of Notts, Ch. Chosen Ochre of Notts, Ch. Corrector of Notts, Ch. Chosen Dinah of Notts, 1948, Ch. Correct Wartax of Notts, 1949. This, too, after she had transferred her affection to the Smooth variety, having previously made the 'Of Notts' prefix equally famous in Wires.

There are several things I will always remember of Her Grace. She was a great personality, respected and admired by everyone. Well I remember her deep contralto laugh, and the immaculate pair of gloves she always wore while showing her dogs on a very, very long, loose lead! No 'propping-up' of terriers for the Duchess!

Another point which comes to my mind: when judging Wires she would pick up her prospective winners, apparently to assess their weight and condition. She did not like terriers of the big, long, weedy type, however long and lean their heads might be! She preferred the cobby, huntsman type of terrier. Legs, feet and coats, too, she considered of great importance. Evidence of this appears in several critiques she wrote for the various club year books.

The late Dr Masters bred six champions, I believe, and Mr Loscoe Bradley, Sir James Hosker and Mr Calvert Butler all succeeded in breeeding five champions each during their lifetime.

Mrs Anthony Blake, Mr Calvert Butler's well-known and popular daughter, has been most successful in producing champions. In 1959 Mrs Blake bred Ch. Watteau Midas. Midas has proved himself a great influence in the success of the 'Watteaus', his progeny having won more than 145 Challenge Certificates. He was sired by Ch. Laurel Wreath, one of Mr Leo Wilson's famous 'L.W.' Smooths. The next champion Mrs Blake bred was Ch. Watteau Lustrous, sire Ch. Full Pay. Lustrous went to Mr Nelson in Johannesburg and did well out there. Watteau Marylyn has proved a wonderful brood bitch; she is by Ch. Hampole Tinkler. Marylyn produced Ch. Watteau Sonata, in 1957, and in her next litter bred Ch. Watteau

Rhapsody and Ch. Watteau Madrigal. These two won both Challenge Certificates at the Fox Terrier Club Show in 1959, and Madrigal also won the Gold Cup.

Next came Ch. Watteau Merrythought and Ch. Foremark Festive (owned by Mrs Newbury), and Ch. Watteau Cantata, born 1959. To bring the list up to date (1964) mention must be made of the lovely tan-marked Ch. Watteau Snuffbox, who, I remember, impressed me very much when I judged him in Scotland in 1962, when he was a raw eight and a half months old puppy. Since then he has gone from strength to strength and has won nineteen Challenge Certificates. His sire is Watteau Sculpture.

As long ago as 1934 Miss Kathleen Emery won the Fox Terrier Club Gold Cup with her Ch. Hermon Conversion Loan. In 1956 Miss Emery again won the coveted Gold Cup at Harrogate with her Ch. Hermond Palmist, again in 1961 with Ch. Hermon Card Trick and again in 1962 with Ch. Hermon Blacklands Sophia.

Miss Emery is the present President of the Fox Terrier Club, and is certainly a wonderful person to know. It gives me great pleasure to mention the successes of some of her favourites.

Pride of place, I feel sure Miss Emery would agree, must go to Ch. Hermon Parthings Loyal Lad bred by Mrs Terrell in 1951, sire Parthings Laddie (litter brother to Ch. Laurel Wreath), dam Amber Solitaire, by Boreham Belucky (Ch. Boreham Belsire).

Ch. Hermon Parthings Loyal Lad has been the winner of seven Challenge Certificates and the sire of Ch. Hermon Palmist, Ch. Hermon Rebel, International and Dutch Ch. Hermon Commando, Ch. Flying Hermon Diamond, Ch. Solus Rosemorder Fire Alarm, Ch. Harkaway Eliza, Ch. Harkaway Emma, Ch. Hampole Sincerity, Swedish Ch. Harkaway Smudge and French Ch. Watteau Loyalist.

Crystal Lady, sire Ch. Hampole Tinkler ex Barrowby Belle, was purchased by Miss Emery to mate to Ch. Hermon Parthings Loyal Lad, and the progeny resulting from these matings have included Ch. Hermon Palmist, Ch. Hermon Rebel, Ch. Hermon Commando and Ch. Flying Hermon Diamond.

Ch. Palmist won eight Challenge Certificates as well as the Gold Cup. She has been most successful, too, with her maternal duties, having bred the lovely Ch. Hermon Card Trick, Irish Ch. Hermon Oracle and Hermon Witchcraft (the dam of Ch. Hermon Fantasy).

The year 1956 saw the birth of Ch. Hermon Rebel, who won

Thurse

Ch. Watteau Chorister (Born 1954)

Ch. Hermon Card Trick (Born 1958)

Thurse

Thurse

Ch. Rawdon Rhoda (Born 1951)

Ch. Madam Moonraker (Born 1953)

Thurse

eleven Challenge Certificates, who sired Ch. Hermon Fantasy, Ch. Shaftmoor Bellechien White Heather and who is the grandsire of Ch. Hermon Blacklands Sophia.

Ch. Hermon Card Trick, by Ch. Burmar Warrior, ex Ch. Hermon Palmist, born in 1958, is the winner of thirteen Challenge Certificates, also winner of the Gold Cup in 1961.

In 1962 Miss Emery won the Gold Cup with her Ch. Hermon Fantasy, bred by Lady Gooch, sire Ch. Hermon Rebel, dam Hermon Witchcraft. I remember giving this bitch her first Challenge Certificate at Cardiff in 1961. She is very lightly marked, as indeed most of Miss Emery's champions have been. She certainly has a wonderful type of Smooth Fox Terrier, and is always to be reckoned with in the show ring today.

One of the earliest champions to win the coveted title after the war, in 1946, was Ch. Lethal Weapon, bred by T. George and sired by Lethean Waters. Ch. Lethal Weapon, one of the famous 'L.W.' Smooths, was owned by Mr Leo Wilson, and proved a great stud force. He sired Ch. Laurel Wreath, Ch. Rush Gleam, Ch. Charnmouth Sea Storm, Ch. Boreham Bendigo, Ch. Farleton Farina, Ch. Flying Larks Wing, Ch. Kingswood Kozy Kole, Ch. Harkaway Lili, Ch. Scroggy Sophocles, Ch. Kenelm Bellechien Pirouitte and, in 1955, Ch. Last O' Weapon.

Of these, Ch. Laurel Wreath, who won the Fox Terrier Club Gold Cup in 1951, was bred by G. E. Hurrel and became the sire of Ch. Maryholm Spun Gold, Ch. Watteau Midas, Ch. Flying Dream, Ch. Lavish War Paint, Ch. Stubbington Matinee Idol, Ch. Watteau Midas.

Ch. Lavish War Paint, breeder F. M. Mills, sire Ch. Laurel Wreath ex Boreham Beguile, was the father of Ch. Correct Wartax of Notts, bred in 1949 by the late Duchess of Newcastle, and which was, I believe, the last champion to be bred by Her Grace.

Ch. Lavish War Paint also sired Ch. Full 'O Pep and Watteau Songstress. From this it will be seen that many of our present-day Smooths own one or other of the 'L.W.'s' among their ancestors.

Although Mr Leo Wilson no longer shows Fox Terriers, he is an all-round international judge, and officiates at many of our largest championship shows, quite often judging Best-in-Show.

In Rev. Dr Rosslyn Bruce's book *The Popular Fox Terrier* he wrote, under the heading 'Valuable Smooth Fox Terriers': 'When a Smooth Fox Terrier is sold at a high price, as when one was recently sold for £1,000, or when two bitches were sold for 800 guineas, the general public is duly impressed, and the newspapers "lap up a bit of a scoop", but what is not so often emphasized is that a considerable number of us have owned terriers, for which similar sums have been offered, but which, quite apart from sentimental reasons, we would not sell for a great deal more. Money is all right, one has no grudge against it, but, after all, its only value is to procure what you most desire, and if you have got that, money loses all value, as far as that is concerned.'

My reason for quoting this is to bring me to the lovely bitch Ch. Gosmore Rosemorder Fireaway, handled by the well-known professional handler Mr F. Taylor to her many victories. She certainly caused quite a stir when she was first shown in 1953, under Mr W. Burrows at Leeds. She was about seven months old at the time, a dainty tan-marked puppy, I remember. Not only did she win all her classes on that day, but also the championship. Quite a feat for so young a puppy. She went on to win twenty-two Challenge Certificates from the Leeds Show where she started to the following Leeds Show the next year; after which she was sold, I am told, for a record four-figure sum, to go to Italy. In spite of all that has been said and written about these very high prices, my experience is that they are exceptional.

Next I would like to mention Herbert Johnson's very famous 'Brooklands' Smooths. Mr H. Johnson is a great supporter of present-day shows. He is a well-known championship show judge, too, and a genial 'gentleman' to know. I purposely use the word 'gentleman' for Herbert, because this is just what he is at all times. Very seldom does one see him not winning top honours, but should he be at the 'wrong end', his genial expression never varies. In 1948 he owned Ch. Brooklands Ebony Girl, bred by S. Jury. This is one of the very few champions carrying the Brooklands prefix not bred by Mr Johnson. Mr Johnson bred Ch. Flying Brooklands Venus in 1948 and sold her to the late Baron Von de Hoop before making her a champion. Thus she carries the famous 'Flying' prefix.

In 1946 Mr Johnson bred Ch. Brooklands Black Narcissus. In 1949 he bred Ch. Brooklands Black Prince, also Ch. Brooklands Happy, who became the dam of Ch. Brooklands Happy Wish. In 1951 he also bred Ch. Brooklands Black Knight.

Although not breeding her himself, Mr Johnson next owned and made Ch. Brooklands Lucky Wishbone, sired by Ch. Brooklands

Black Prince and, according to the Kennel Club Stud Books, bred by Mrs L. C. Wilson. Also in 1951 Mr Johnson bred Ch. Brooklands Black Mask, by Ch. Brooklands Black Prince. Next came, in 1952, Ch. Brooklands Black Ace, again bred by his owner, and sired by Ch. Brooklands Lucky Wishbone.

In 1953 appeared Ch. Brooklands Lucky Dip, bred by Mr Johnson, sire Ch. Brooklands Lucky Wishbone ex the Champion bitch Ch. Brooklands Black Narcissus. This was the second champion bred from Brooklands champion bitches.

We still have not finished with H. Johnson's home-bred champions, for in 1956 he bred Ch. Brooklands Royal Tan and Ch. Brooklands Happy Wish. Ch. Brooklands Happy Wish is the father of Ch. Brooklands Merry Thought. In all, Mr Johnson has actually bred eleven champions. Besides this amazing achievement he has owned, shown and exported too many beautiful Smooth Fox Terriers to list. However, I feel that I cannot omit to mention and congratulate Mr Johnson on the latest Brooklands star, Ch. Brooklands Present, bred by Mrs Blake and sired by her Ch. Watteau Chorister.

This brief outline of the Brooklands Smooths clearly shows the great influence this world-famous prefix has on the breed today. It also illustrates the basic principle of breeding only from the best bitches obtainable.

Mr J. Lowe, writing in the Fox Terrier Club Year Book of 1961, says just what I have been trying to say for years, and with his permission I quote: 'If you wish to breed high-quality dogs you should plan your matings with the sole object of breeding tip-top bitches. The old saying that a kennel is just as strong as the bitches it produces is not just a cliché but a statement of fact. If and when a bitch puppy is bred better than her dam, retain her, and in course of time breed from her and repeat the whole selective process. In this way, your high-quality young dogs will appear from time to time. It has been proved again and again, that the people who breed to improve their bitches are usually those who bring into the show ring, year after year, high-class dogs who more than hold their own in the hottest competition and obtain a high proportion of the top awards.'

Another matter often noted when perusing pedigrees is the overwhelming number that show the bottom line, i.e. the dams' line, to be the weakest. The goal for all intelligent breeders should be to make this part of the pedigree the strongest line.

It gives me great pleasure to publish a short history of the successes of Mr and Mrs John Lowe's 'Lanneau' Smooth Fox Terriers. Both Mr and Mrs Lowe are well-known championship show judges of both Wire and Smooth terriers, and also authoritative judges of all terriers. Mr Lowe is the Secretary of the Smooth Fox Terrier Club, and he has given me permission to publish some of his 'Thoughts at Random' from his year books.

Writing about colour, he states: 'For some time now, black and whites appear to be on the increase in our breed. This is all to the good. Colour can, and does, play an important part in a breed's popularity. There is an old Fox Terrier saying, and I think a true one, that an outstanding black and white terrier is hard to breed, but when one is lucky enough to possess such a treasure, you usually get a better head and more varminty expression. Another notable feature, which black and whites carry on their forelegs, is the old Huntly spots, named after that great hunting man, the late Marquis of Huntly. One of my most valued possessions is a letter from the late Mr James M. Austin, in which he comments on the beautiful black and white markings of his great dog Nornay Saddler. This dog also carried the Huntly spots, mostly on his forelegs. Mr Austin asserted that his glorious colouring, plus these spots, helped considerably in his dog's many hard-won victories. Clear black and whites, pure tan and whites, and all-white Smooths are the colours we should aim for in our stock. . . . Dirty tans and brindles, if they are used in our breeding plans, can, and do, bring problems in colour breeding which on occasions are almost insurmountable.'

Ch. Lanneau Jewel, born in 1947, by Firstmonsieur ex Lanneau Victoria, bred by J. Lowe, was the first Lanneau champion. The year 1949 was the period of her great successes. Between March and September of that year she won nine Challenge Certificates and was almost unbeatable, winning the Duchess of Newcastle Challenge Cup for the Fox Terrier (Wire or Smooth) winning the most Challenge Certificates in the year. To add lustre to her achievements in that year she also whelped her first litter. From the puppies born from this litter all the show successes of the Lanneau Kennels up to date are directly attributable.

I am told that Jewel had an uncanny ability for timing and perfection in showing, and in this respect Fox Terrier people to this day comment on this great aid to her show career. She was always shown on a loose lead and resented most stubbornly a tight one.

Jewel developed the habit of refusing to leave the ring after winning at a show until she was allowed to walk round the ring to receive the pats and congratulations of the ringside! In her kennels at home she was a law unto herself. She ruled the bitches from the youngest to the oldest with a rod of iron! Conversely, none of the male dogs could do wrong in her eyes. She died at the age of fourteen years, a great character indeed.

Ch. Lanneau Jerrod (1952) was to become the next Lanneau champion, and in 1954 Mr and Mrs Lowe bred Ch. Lanneau Jeremy, sirei Ch. Lanneau Jerrod, dam Lanneau Jenta. The sire Ch. Jeremy was a great grandson of Ch. Lanneau Jewel on the dam's side.

Ch. Lanneau Jeremy won five Challenge Certificates in 1956. This partcular year, I believe, was a very difficult one so far as Smooths were concerned, there being five or six outstanding dogs 'fighting it out', who all subsequently became worthy champions.

Jeremy was exported to Rhodesia in 1957 and, from reports received, has, through his stud work, brought popularity to the Smooth Kennels in that country.

In 1956 Mr and Mrs Lowe bred Ch. Lanneau Jessica, sire Ch. Brooklands Black Ace, the dam being, again, Lanneau Jenta.

To bring this list up to date comes Ch. Lanneau Jekyll, born in 1960 (breeder/owners), sire Lanneau Jeweller (Lanneau Jenta), dam Lanneau Jezebel (Ch. Lanneau Jessica). Jekyll became a champion in 1962, winning four Challenge Certificates in that year.

While going through the records of the Smooth champions I have come across the name of Ch. Selecta Rich Reward several times. I remember judging him in the West Country when he was quite a young puppy, not more than seven months of age, if I remember correctly. He pleased me very much, and his owner seemed delighted when he won several mixed classes of both Wires and Smooths. He was a rich tan-marked dog, sired by Golden Spur of Sker ex a bitch called Lucky Strike, and bred by P. Davenport. Later Rich Reward was purchased by Mr H. R. Bishop, who added his well-known prefix of Selecta to the original name. In 1949, when he was about five years old, he became Ch. Selecta Rich Reward. He has sired the following champions: Ch. Full Pay, Ch. Black Andrew, Ch. Sheresta Monogram and Ch. Touchwood Tribute. Ch. Full Pay became the father of Ch. Watteau Lustrous. Ch. Black Andrew sired Ch. Sheresta Miss Andrew and Ch. Sheresta Monogram sired Ch. Sheresta

Borman Vista. I am very pleased to find that Ch. Selecta Rich Reward has had such an influence in producing present-day champions.

Mr John Glover, of the Hewshott Smooth Fox Terriers, seems to have bred a great many good terriers with unfailing regularity for a very long time. He had bred Ch. Hewshott Don Juan, and in 1949 he bred two champion bitches in the same litter. Ch. Hewshott Joyful Light and Ch. Hewshott Joyful Lark. Ch. Hewshott Jaguar came in 1953 and, in 1956, Ch. Hewshott Juggler, sired by Ch. Hewshott Jaguar. In 1963 Mr Glover was again successful with Ch. Hewshott Jennie Jerome, which he bred from Hewshott stock.

Mr A. Clanachan has bred and owned many champions. His first home-bred champion was born in 1946: namely, Ch. Maryholm Silver Lady. Ch. Maryholm Spun Gold (1948), Ch. Maryholm Simon (1949), Ch. Maryholm Sweetmeat (1954), Ch. Maryholm Sailaway (1955) and Ch. Maryholm Sureline (1958) were all bred by Mr Clanachan. Next came Ch. Maryholm Nornay Mainsail (1959), bred by Messrs Coward & Soubrey. Also in 1959 came Ch. Maryholm Royan Surepay, bred by J. Morton, and in 1961 Mr Clanachan bred Ch. Maryholm Ship Ahoy. Lastly, in 1963, Mr Clanachan made up Ch. Maryholm Silver Snowflake, which he bred from his Ch. Maryholm Sureline ex his Ch. Maryholm Silver Lady. This must be regarded as quite an outstanding record.

I realize that there are many other astute breeders who have each bred several champions, but there are too many of them to give in detail here. The appendix shows the post-war champions up to date. The Kennels I have particularly mentioned are owned by exhibitors I have known and met personally over the years, which reminds me of two more: Miss Stapley, who has bred four champions, and Miss Linda Beak, who has bred her first champion Smooth in Newmaidley Jehu. Jehu was born in 1960 and I have no doubt this is only the beginning of Miss Beak's Smooth champions.

I have to admit that my principal experience has been with the wire-coated Fox Terrier, but I have always had a fondness for the smooth variety also. There is, to me, a special appeal about Smooth Fox Terriers. I particularly admire their clean-cut outline, their lovely chiselled heads, their beautiful legs and feet, and their gay, arrogant carriage. A good Smooth is a handsome, well-tailored terrier, which always attracts attention.

CHAPTER 6

THE BROOD BITCH: WHELPING

I do not propose to go into this all-important subject from the veterinary surgeon's point of view, but rather as a guide to people breeding their first litter of puppies. This is a most exciting and wonderful experience. Many people are nervous at this eventful time, quite naturally so, if they have never been present at a birth and have read harrowing articles of things that may, and probably will, go wrong. Very rarely does one have a bad whelping with Fox Terriers, which are generally easy whelpers and excellent mothers. Little actual assistance is needed, but common-sense supervision is necessary.

Many factors must be taken into consideration when deciding the most desirable time to mate a young bitch. Fox Terriers sometimes come in season as early as seven months of age. This, of course, is far too early to consider mating them. Eighteen months of age is almost perfect, I would say. This would be at their second heat.

If one is campaigning a bitch with the hope of her becoming a champion, quite often even the second heat has to be missed. I do not like mating very young bitches for the simple reason that they are not fully matured. Should one have a bitch, say over twelve months of age, which has a 'shelly body' and is a bad feeder, however, it might be as well to let her have a litter with the idea of her filling out and feeding better. Once a bitch has had a litter, it is surprising how much more eager she is for food, bodying up more quickly in consequence.

It is usual for bitches to come in season (heat or oestrum) twice during the year. Many bitches, on the other hand, only come in once in eighteen months. This is quite normal, so do not be alarmed if this should happen to your bitch. I have even had them go as long as twelve months between their seasons.

The season usually lasts for about three weeks, and naturally they must be carefully guarded so that they cannot come in contact with any other dogs. It is wise to make a note, with the first heat of a bitch, just how long it lasts, so that for future occasions you will know which will be the correct day on which your bitch must be mated.

The first signs of a bitch coming in season are a distinct swelling and pinkness of the vulva, then follows a red discharge, which usually increases for the next seven to ten days. It is very difficult to advise which day is the best to have your bitch mated because they vary so much in the length of their season. Usually the 12th to the 14th day is about right. On the other hand, I have known bitches that have to be mated as early as the 5th day. Experience alone of your bitch will give you the right answer. It is generally considered that the best time is when the coloured discharge has faded and practically gone.

Having decided to have a litter from your bitch, immediately she starts in season write or telephone the owner of the stud dog you have chosen, booking a service to the dog you have selected as the most desirable mate for your bitch. Do not let distance decide your preference for a dog, but choose one that has proved his capabilities of producing strong healthy litters, with a winner or two to his credit as well.

If you are able to take your bitch to be mated so much the better, but if the dog is a long way away it is much better to send your bitch rather than to use a dog near at home, which may not have a suitable pedigree at all.

When sending a bitch by train she should have a good, strong, well-ventilated box in which to travel. This must be large enough for her to stand up in, and to be able to turn round, and so settle herself comfortably for the journey. Two clearly written labels, giving the name and address of the Kennel to which she is going, must be fixed on the box. Also, give the telepone number, so that the station clerk can telephone as soon as she arrives at her destination. I find that owners of stud dogs are kindly people and invariably arrange for the bitches to be fetched from the stations promptly. The bitch's pet name should also be written on the label, since use of the name with which she is familiar helps her to feel more at home while among strangers. Give her a good bed, and be most careful to ensure that she is safely fastened in. I myself always use a padlock and key, tying the key securely outside the box.

Next, find out from your local station the time of the best train for her to travel by. I always send mine by train overnight. They travel much better at night, spending most of the time sleeping, and their arrival early in the morning gives the owner of the stud dog ample time to attend to her comforts and to give her a chance to settle down before she is mated. If she is going by train I suggest it is best to send her on about the 10th day of her heat, so that she can have, perhaps, twenty-four hours' rest after her journey, before mating.

Stud fees, plus return carriage, are payable at the time of mating. Some stud-dog owners allow their dogs to be used for 'pick of litter', but in my opinion it is much better to pay the fee, otherwise you will most probably lose your best puppy, or if there should be only one puppy you may be compelled to hand that over.

The mated bitch should be allowed some hours of rest before she is subjected to the return journey. The owner of the stud dog should inform you when she is to be despatched so that you know when to expect her.

I always worm my bitches before they are mated. They may also be wormed, if necessary, up to three weeks after mating. It is not wise to do this later in the pregnancy.

The actual mating of the bitch is a matter of experience. No two owners of stud dogs handle this most important operation similarly. Some dogs are very difficult to manage, and even have to be muzzled to prevent their turning on the bitch. Most bitches have necessarily to be muzzled upon being introduced to a strange dog. They are naturally nervous at first, but soon settle down if the two terriers are allowed time in which to make friends. This is the essence of the business. In nature the bitch would be courted for a long period, sometimes over a number of days, by the dog before she will allow herself to be mated. It is obviously foolish therefore to expect a satisfactory mating to be over in five minutes. Apart from this, nature has provided that when the bitch is mated the two animals remain tied for a considerable period of time. Some breeders believe that a lengthy tie is essential if a good litter is to result. My experience varies in this respect, but I agree that a good tie is desirable. Care should be taken, however, to ensure that no harm or damage from the tie should occur to either dog or bitch. Patience is essential. Sooner or later the two terriers will become disengaged.

My advice to the novice with regard to stud work is to take every opportunity to be present at a mating. Only experience will teach you the best method to adopt for your dog or bitch.

Should by carelessness or accident your bitch escape from the kennel and become mated by a strange dog do not despair that she has been ruined for life! The best course is to take the bitch immediately to your veterinary surgeon, who will, if you wish, give her an injection to prevent her having the cross-bred litter. In my opinion it is no good mating the bitch to the dog of your choice after she has had such an injection. Far better to miss her for that heat and to wait until she comes into season again. The fact that she may have been mated incorrectly previously, or even may have had cross-bred puppies, will have no bearing upon her producing a pure-bred litter later. Some old-time breeders believe that the subsequent litters after a *mésalliance* will be tainted. This is quite incorrect.

The gestation period is sixty-three days. This varies a great deal. Quite often the puppies are born a few days earlier, or one or two days late. My experience has been that if a bitch is really heavy—that is, carrying a number of strong active puppies, say five or six, which is considered a big litter for Fox Terriers—she usually has them early rather than late. This is because they are a discomfort to her, and her inclination is to whelp as soon as possible. Should there be only one or two bigger, more lazy, puppies, the whelping is more often late. A general guide is given in the whelping table (Appendix E).

During the first four weeks of pregnancy it is advisable not to alter the usual routine of the bitch at all, letting her have her usual exercise and good food. At the same time she should not be allowed to run up and down stairs or to jump. Fresh water should be at hand the whole time for her to drink. At about the fourth week after mating she will probably refuse ordinary kennel food, and then her likes and dislikes should be considered. Each bitch has her own fancies at this time. Sometimes a little grated cheese on top of the meat or cooked rabbit will tempt her to eat, but one should be careful to take out all bones, which splinter very quickly and would cause trouble. After a few days it will be found that the bitch will come back to eating her usual food—that is, plenty of meat, either cooked or raw, with a little brown bread or biscuit meal soaked in gravy as bulk. Instead of the one main meal she should now have two good meals a day, together with drinks of warm milk in between. Fish may be given occasionally for a change. Also raw eggs beaten up in milk. It is not necessary to boil the milk for an in-whelp bitch, but it must be given warm. Boiling the milk is to prevent diarrhoea, but this is not usually necessary for the mother-to-be, who is more likely to be constipated. Six to eight Vetzyme tablets a day provide the minerals and vitamins so necessary for the proper development of the puppies.

A transparent discharge from the vulva about five to six weeks after mating is an almost certain sign of the bitch being in whelp. This should be wiped away with cotton wool, since if the mother is very heavy with pups she cannot reach round to lick herself clean.

Between the seventh and eighth week of the pregnancy the hair should be carefully cut from under the abdomen of the bitch, to enable the newly born pups to get to the vital milk supply; also under the tail and around the hind quarters so that she can easily be kept clean.

Now about the place for the puppies to be born. If there is a spare room in the house or a building adjoining the house which is well heated, and with lighting, this would be quite the best place for the whelping. The pups, I find, are nearly always born during the night or in the early hours of the morning and it is much easier for the attendant not to have to run in and out all night or to spend the night outside.

Warmth, to my mind, is absolutely essential. I do not think the room can be kept too warm for the first forty-eight hours. One must remember that the whelps are coming from a very warm place into a cold hard world; I am sure that more new-born puppies die from being chilled than from any other reason. Even after the first forty-eight hours, and when the puppies have been dried and cleaned, it is still necessary to keep the whelping box very warm, at least 70°F., so that the small quantities of food which the new-born puppies take at first from the mother is used for growth and development rather than for the maintenance of body heat.

A well-made large box is ideal for the bitch to have her puppies in, see Figs. 10 and 11. It should have a lid and front to open, so that the attendant has easy access to the dam, and for easy cleaning out. It is preferable that the box should be of such a depth as to be most comfortable for the attendant to bend into when necessary.

I advise a closed-in box because should the bitch whelp without anyone in attendance she may well scatter the puppies in different parts of the room or kennel, with consequent certainty of their becoming chilled, and risk of death. This is especially so if the bitch has been trained to be house-clean, as she would naturally become apprehensive of punishment for being dirty, and would move from the soiled parts of the floor. Also, it is much easier to keep a box of the size illustrated warm, than a whole room or kennel.

I realize the difficulty of the bitch not being able to run in and out to relieve herself, but by the time she really takes to her bed she will not need to do so for some twenty-four hours at least. This is because before she starts straining in earnest to give birth to the first puppy she will become uncomfortable and will pass urine several times and evacuate all excreta.

A completely closed-in box with a wire-mesh front, which can be covered over with a blanket as soon as the bitch has settled down with her new-born family, is best. They feel much safer and warmer closed in.

For the actual whelping I use clean, dry newspaper to which more can be added on top after each puppy has been born. It does not stick to the puppy and is something the mother can scratch and tear up without causing her harm. Straw is not so absorbent, and hay, too, becomes very messy. Hay and straw not only stick to the puppies while they are wet, but if the bitch has the puppies quickly, should she be alone, they can very easily become buried in the straw or hay and quite likely squashed by the mother while she is still in labour.

I also have ready two large, white cotton pillow-cases filled with woodwool, or newspaper screwed up, which I make to fit the floor of the whelping box. In addition, several pieces of old, clean white blanket are required. When all the puppies have been born the dirty paper can be taken away for burning, and the filled pillow-case can be placed flat on the floor of the box and the blanket tucked well in all round. This makes a lovely clean, warm bed for the mother and family.

The whelping box should be ready and in position at least a week before the puppies are due, so that the bitch may sleep in it at night, with plenty of newspaper for her bed.

The normal temperature of a dog is 101·8°F., but about twenty-four hours before the puppies are born the bitch's temperature will be found to be subnormal; that is, about 100°F., and will fall even lower just before actually whelping. The method of taking a dog's temperature is to insert the end of a veterinary thermometer, which

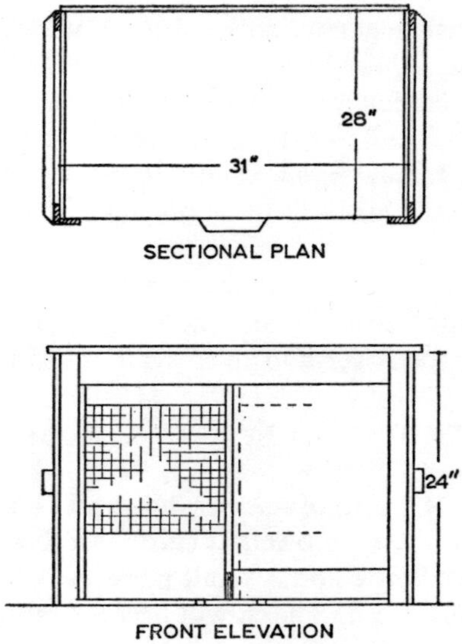

Fig. 10 Timber whelping box

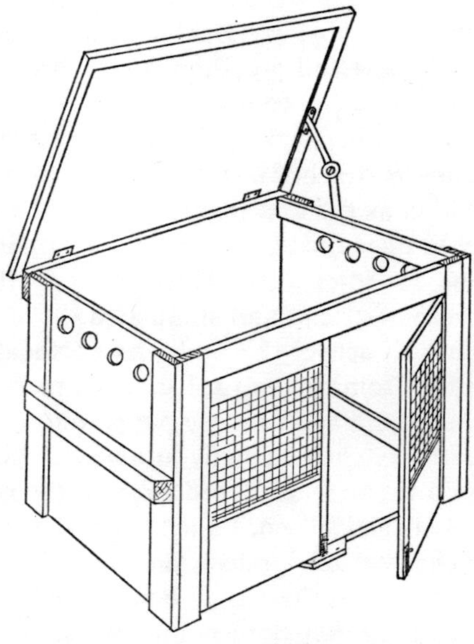

Fig. 11 Timber whelping box

has previously been smeared lightly with disinfectant ointment, into the rectum for the required period.

Many bitches, especially with their first litter, become restless, uncomfortable and unhappy-looking for about twenty-four hours before the actual labour pains begin. They may refuse their food, but not always, and this is little or no indication that whelping is imminent.

The usual sign that the preliminary pains have started is that the bitch will pant quite frequently and shiver violently. She will tear and scratch the newspaper over and over again, making and re-making her bed.

Now is the time to collect the few things necessary during the whelping. They are:

A bottle of Dettol, a pair of scissors, hot-water bottle, cotton wool, several pieces of white rag, towels, a cardboard box large enough to take the hot-water bottle and a small piece of blanket.

As the pains increase, the bitch will become more and more restless. Eventually she will really start to strain hard, not all the time, but in short, sharp bouts, until the first puppy is born. If the period of actual straining continues for more than two hours it is advisable to inform the veterinary surgeon. Great patience is required with the whelping bitch; calmness and precision is very necessary in all that has to be done to help and encourage her.

Each puppy is beautifully packed in its little transparent container, enclosed in fluid, and is attached by the umbilical cord to the placenta (after-birth). As soon as the first puppy is born, the mother, in her natural state, breaks open the bag, by tearing it with her teeth, severs the cord and eats the after-birth. Then she licks the puppy continually until it becomes clean and stimulated.

I help my bitches to deliver their puppies personally. As soon as the puppy protrudes from the vulva, I take a small square piece of white rag and gently hold and ease the puppy out with a downward movement while the bitch is actually straining. Then, still with a piece of rag, I hold the umbilical cord close to the puppy so that I can pull out the after-birth. Then, I quickly break the bag open near the mouth of the puppy and wipe away any fluid that may be around the mouth. I do this first, as there is no immediate hurry to attend to the umbilical cord. The newly born puppy will soon appear to gasp for breath, which is an indication that all is going well. Now I cut the umbilical cord about two inches away from the puppy with a pair of scissors and then return the puppy to the mother. The after-birth is put into a bucket to be burnt later.

While all this is going on, the bitch will most likely be busy licking herself and cleaning up the fluid. If she resents the handling of the puppy she must either be left to attend to it herself or the operations I have described can be done while the puppy is still in the box with the mother.

Some breeders think it best for the bitch to be allowed to eat the after-birth, and for the beginner I suppose it is much easier to 'leave it to nature', but I find that eating the after-birth sometimes makes the bitch sick if eaten too quickly. It also causes looseness of the bowel. Also, from my experience, I find that maiden bitches take quite a long time to understand just what to do with the first puppy or two, and the puppy can die if not released from the bag of fluid quickly.

The beginner may decide to leave the mother to sever the umbilical cord herself, and it will be found that after she has actually severed the cord with her back teeth, she will then draw the piece that she has left attached to the puppy through her front teeth. There is no reason to fear that this will hurt the puppy, as this is just nature's way of preventing any excessive bleeding. The remaining piece of cord will quickly dry and fall off in a day or two. Even if the dam attends to the severing of the cord herself I still think it helps in giving the puppy a good start in life if the fluid is wiped away from the mouth until it breathes properly.

The time between the birth of each puppy varies a great deal. Sometimes two are born within a few minutes of each other. More often, there is a gap of anything from twenty minutes to one and a half hours between the birth of each puppy. During the breaks between the pains the bitch will often curl up and let the puppy snuggle up to her. Quite often, the tiny damp new-born puppy will begin to suckle, but as soon as the strong labour pains start again they are usually too great for the mother to bother about the first-born. When the birth of the second puppy is obviously imminent I take the first one out of the whelping box and put it into the card-board box on top of a covered hot-water bottle and place a small piece of blanket over the top of it. This soon makes the puppy warm and dry, and stops it crying. This is important, because if the dam hears the puppy crying she will become restless and will not attend to the birth of the next puppy. Here common sense must prevail, since, if the mother resents the puppy being taken away, it must be left with her.

However, if possible, it is much preferable to dry the puppy off on a bottle. Should the puppy be left with the mother during the whole time of the whelping, and until all the puppies are born, quite probably it will be dirty and bloody, as well as damp.

Leave the puppies with the mother until she has settled down and obviously finished whelping, and then it is advisable to take the puppies out one at a time and wash them gently with cotton wool dipped in hot water. The puppy should be held in one hand and stroked with the cotton wool, and dried as much as possible with a soft towel.

When all the puppies have been born the mother will settle down quietly to sleep and the whelps will start to suckle. Care should be taken to ensure that the puppies are feeding before the attendant leaves the whelping quarters. It is very important that each puppy should have its fill of the mother's first milk, known as the colostrum, which nature provides with a special composition to meet the needs of the new arrival. The colostrum is thicker and less fluid than the subsequent flow of milk, and in consequence the initial food supply appears to be comparatively small. As the suckling continues, the flow increases. It is a matter of supply and demand. The more the puppies take, the more the milk glands will fill up. Conversely, if the glands are not utilized they soon dry up. The bitch should be given a drink of warm milk and left alone for about an hour, with the front of the whelping box covered with a blanket, after which all the wet, dirty newspaper should be removed. Do not be alarmed at any greenish-coloured mess in the box, this is quite natural. The prepared mattress should be placed in the whelping box covered with a blanket. The soiled hind quarters and tail of the bitch should be quickly sponged before returning her to the puppies, so that they may be near the life-giving warm mother.

It will soon be apparent whether or not the new-born puppies are feeding and doing well. If they cry at all it is usually because they are, for some reason or another, not getting milk. There may be one small puppy that is being pushed away by its bigger, stronger brothers and sisters. If this happens, and the starving puppy becomes a 'squeaker', something must be done about it, as it upsets the mother

Wire Fox Terrier puppies at six weeks

Smooth Fox Terrier puppies at eight weeks

Wires at Leicester Championship Show, 1963
L to R: Ch. Gosmore Kirkmoor Tessa, Ch. Brooklands Elegance, Ch. Seedfield Ernley Empress, Penda Deko Delmage, Penda Silver Token

The author showing Ch. Penda Peach

F. W. Simms

and makes her unhappy and restless. I find the best thing to do is to hold such a puppy on to one of the mother's teats every two hours for the first forty-eight hours, making certain that the puppy actually gets some milk, and is not sucking in vain. This often happens if the others have been there first.

After the first twenty-four hours it will be obvious which of the puppies are doing well and which, if any, are looking flat-sided as compared with the others. It is easy to tell which of the puppies are doing well by placing a puppy flat on the hand. If it wriggles about strongly we say it has a lot of 'curl' in it, and are well pleased. If the litter is uneven in size and strength the two or three fattest, strongest puppies should be taken right away from the mother and put in the cardboard box under a small piece of blanket, with a warm, not hot, hot-water bottle, for about an hour, so that the smaller puppies can suckle in peace while the 'Billy Bunters' are away.

The 'piece of blanket' tip is well worth remembering. Whenever the bitch is taken away from the puppies, to relieve herself for instance, I always completely cover the babies with a warm, light, clean piece of blanket. They will not suffocate, and it will definitely stop them crying. This is important, because if the mother hears them she will be frantic to get back to them and even be clumsy and careless in her endeavour to get back quickly, quite often hurting them by rushing in and even jumping on top of them.

Fox Terrier puppies are born with good, strong, hairy coats. The markings are very distinctly defined, usually black, with slight tan on the heads, which turn completely tan when the puppies are between six to eight months old, after losing their puppy coats. As soon as the puppies are dry, the coats look beautifully white against their black or tan body markings.

Noses are always pink at birth. They gradually change to black, and are usually quite black at about eight weeks of age. Occasionally, some are mottled. These, too, usually turn completely black, but take much longer to do so, as long as eighteen months sometimes.

I mention this fact because on occasion people breeding their first litter of Wires have telephoned me saying, 'The puppies are no good, they have black heads and pink noses!' I have put their minds at rest by assuring them that in ninety-nine cases out of a hundred the noses turn black.

Their eyes are completely closed at birth, and gradually open when they are about twelve to fourteen days old. Their ears, too, are closed, and they cannot hear for the first three or four weeks of their young lives. While the ears are closed there is no 'life' in them. By this I mean that the puppy cannot lift or use its ears, and therefore one cannot tell how the ears will be carried in later life.

Perhaps the reason that the eyes and ears are closed at birth, and for a short time afterwards, is that the natural place in which the bitch would choose to have her puppies would be in an earth dug by the mother, with danger of damage to the organs from the soil and grit. By the time the eyes and ears are open the puppies would be strong enough to move and hold up their heads.

The bitch will be loath to leave the puppies for the first two or three days, therefore she should be fed in the whelping box, and taken out just long enough to relieve herself, say twice a day.

For the first forty-eight hours after the last puppy has been born the bitch should have a very light diet—a little and often, is by far the best. She should be given warm milk or Lactol, Farex made with warm milk, or beaten-up egg and warm milk, gradually getting her on to more solid food. This may consist of brown bread and warm milk, cornflakes and warm milk, then perhaps a little steamed fish or a little rabbit added to the brown bread which has been previously soaked with gravy or warm milk and mashed small. Sugar, in my opinion, is not advisable, as it often causes diarrhoea.

After the first two days the bitch should be given all she will eat of her usual diet, with several drinks of warm milk in between. If the bitch should become sick of milky foods at this time such things as Bournvita or Ovaltine may be added to the milk to make a change.

The tails of the puppies when born are too long for present-day requirements. It is therefore necessary to dock them to the required length. This should be done at about seven days after birth. Some people do this at the third or fourth day, but I think the seventh or eighth day the better, for two reasons: one, as the days go by, the mother will be happier spending a little more time away from her family; and, two, the whelps are stronger to withstand the operation.

This is my method. A sharp pair of scissors, a pair of curved nail scissors, permanganate of potash crystals and cotton wool are required. First, the two pairs of scissors are sterilized by boiling in water. They should be taken out of the dish in which they have been boiled with a fork or a pair of cooking pincers, and placed on a clean piece of towelling or blanket on top of the whelping box or table. Grind the permanganate of potash very finely with a pestle, or it can be done with a teaspoon. Each tail is considered carefully and individually, remembering that they vary considerably in length and thickness. The correct length of tail can greatly improve the appearance of a Wire or Smooth Fox Terrier, and a short dock completely spoils the picture. Usually, about one-third of the tail should be taken off, but in some cases when the puppy is a thick-set cobby type, with lots of bone and a thick tail, it is necessary to take off only the tip. With a more racy type—that is, one with lighter bone, longer in back and tail, which invariably go together—the one-third rule is about correct. Before even starting to do the actual docking I take each puppy out of the box, place it on a blanket on the table, hold the head and tail up, and mark with an iodine swab, consisting of cotton wool on a match, the place at which I think it will be correct to cut.

I have been thought a fussy old woman about this, but, believe me, I have seen many puppies completely ruined for show by being docked too short. Remember, one can always have a piece taken off later by a veterinary surgeon, but to be able to stick a piece on, never! However much hair is later grown on the tip of the tail to hide its shortness, the tail never looks as well as a good strong dock of the right length. To see the hair blowing in the breeze on top of the tail is an abomination. So remember, it is better to be safe than sorry, and to err on the long side rather than the short.

Next, take the mother and put her out in her usual exercising place and proceed with the docking.

It is a great help to have someone to assist. Ask the helper to place his hand over the front half of the puppy to keep the puppy still. Now, with the left hand, hold the tail up, have one final look at the length, and, while pressing the skin down as much as possible towards the base of the tail, cut off the top of the tail in one sharp, quick cut. The puppy hardly squeaks at all—not so much as when the mother accidentally stands on it. The tail will begin to bleed slightly, so quickly dab on the permanganate of potash powder with a piece of cotton wool, using it liberally, thus completely sealing the wound. No blood should be lost at all. Having tried many so-called certain 'blood stoppers' over the years, I find permanganate of potash quite the best of them all. Having done this, cut all the little toenails, being careful not to make them bleed by cutting the quick. This is the little pink line in the nail. If left uncut the nails grow very quickly into sharp needles, which scratch the mother while the puppies are suckling and so make her unsettled and uncomfortable.

All Wire Fox Terrier puppies are born with dew claws, or thumbs, on their front legs, and these should be removed. This should be done at the same time as the tails are docked. At this age the dew claws are no more than gristle, and can easily be removed with a sharp pair of curved nail scissors, previously sterilized. Press the scissors under the nail joint, close to the leg, and with a quick movement cut the whole small joint off. As soon as it starts bleeding, dab on the permanganate of potash, as previously described for the tail operation.

Having docked the tails, taken off the dew claws and cut the toenails, return the puppies one by one, covering them over with a piece of blanket until they have all been treated, and then return the mother to them. Their troubles are all over for the time being, and they should sleep, eat and grow with astonishing rapidity, with the mother to keep them warm and clean, and in constant attendance.

Seldom does one have to call in the veterinary surgeon to a whelping bitch, but should the bitch show no signs of having the puppies after two or three days over her time, it is as well to seek his advice.

Should there be a black discharge from the vagina before whelping, call your veterinary surgeon at once, as this means trouble, and is probably an indication that there may be a dead puppy, and in the case of a maiden bitch she may have trouble in giving birth to it. This is mentioned not to frighten inexperienced breeders but rather as a warning.

It may allay apprehension if I tell you that in my forty years of breeding Wire Fox Terriers I have never lost a bitch through whelping.

Be careful and watchful, but not nervous.

CHAPTER 7

PUPPIES

FROM the events described in the previous chapter, onwards, the puppies should spend all their time eating, sleeping and growing. When being handled they should feel firm and their tummies round and hard. They gradually become more active, and at about four weeks old they are usually well up on their legs, and moving about the whelping box quite quickly.

At about this time, particularly if a big litter, they begin to look a little less fat, and this is the time to start supplementing the mother's milk. The first meal they have should be a very small quantity of scraped lean beef. About half a teaspoonful is quite sufficient for each puppy. It should be scraped with a spoon, since this avoids the risk of including any sinews; scraping the meat renders it much more digestible.

Each puppy must be fed separately. The best way is to hold a small quantity of the meat between the finger and thumb. The first time or two, push the puppy's nose into it; he will soon appreciate the taste and take the meat rapidly. On the second day each puppy may have two such feeds a day.

When the puppies are about four weeks old it is a good idea to give the mother two extra drinks a day of either Lactol or any proprietary baby-milk food. The whelps will now be demanding more and more milk from the dam and I advise giving the bitch the same milk food as the puppies are going to have.

On the third day of the supplementary feeding the puppies may be given a small quantity of such milk food, fixed with *boiled* water, as directed on the label—no sugar. As I have said before, sugar is inclined to cause looseness of the bowel. Again, each puppy must be introduced to the milk food separately. A saucer is ideal for their first feed or two. Some take to lapping much more quickly than others, but usually after the first day or two of blowing bubbles into the food they feed quickly and well.

Should this extra food give the babies diarrhoea it is advisable to stop supplementing their diet for about twenty-four hours. If diarrhoea is allowed to continue for any length of time the puppies become very sore around the rectum and under the tail. It must be remembered that puppies of one month of age are babies, and must be treated as such. Clean under the tail with cotton wool dipped in olive oil, and then sprinkle healing toilet powder on to the affected parts. Some people use only olive oil of 'Vaseline' for this purpose, but I find that the mother then licks the puppies to make them clean and so makes their little sit-upons sore.

At five weeks of age the milk food may be thickened with Farex for one meal a day. Continue with the scraped raw meat for two meals a day, gradually increasing the quantities.

The next step towards complete weaning is to introduce brown breadcrumbs into the milk food. Just a little at first, gradually increasing the quantities, each puppy being fed separately and its requirements studied individually. Puppies usually vary quite a bit in size and appetites.

By this time the mother will be spending more and more time away from her family, and she should now be fed away from the puppies, because if they try to eat her food she may snap at them. My advice is to feed her in a separate kennel, right away from her offspring, and not to return her for at least one hour afterwards.

Some terriers have such a strong maternal instinct that if the bitch is left with the puppies or returned to them too soon they will regurgitate their food for them. This should not be allowed to happen as it tends to pull the bitch down in condition. It is much better for the puppies to have the correct quantities of the food as described.

At six weeks of age the puppies should be having four meals a day, together with any milk the mother can now offer them. Some bitches will continue to feed the puppies for well after eight weeks, while others become tired of them much earlier. There is no hard-and-fast rule that can be applied. It is best to allow each bitch to decide for herself when she wishes to leave her family.

The only other addition for puppies of between six to eight weeks old is a very small quantity of wholemeal puppy biscuit, previously soaked in boiling milk or boiling water, and added to the food. Cooked, finely minced meat should now be added to either the soaked brown breadcrumbs or to the soaked biscuit meal.

All food should be fed in a sloppy state to puppies of this age, not just moist or stodgy, which they find difficult to swallow.

When the puppies are about four weeks old the floor of half the whelping box is covered with newspaper and the other half made into a bed with blankets. The puppies soon learn to use the newspaper to relieve themselves. At six weeks of age they have clean sawdust on the floor of the whelping box or kennel and are allowed to come out of the whelping box for short runs.

If the puppies have their food given to them in feeding bowls on top of the sawdust it is advisable to spread a blanket under the bowls. Care should be taken to wipe their faces after each meal. Sawdust sticking to their whiskers and being eaten does not help.

At eight weeks of age the puppies should be still having four meals a day and, in addition, if winter puppies, a hot drink last thing at night. Two of these meals should be meat and biscuit meal, and the other two of milky food such as Farex, cornflakes with boiled milk, served warm, or egg beaten up with milk. One egg is sufficient for the whole litter. I do not care for vegetables in any shape or form for dogs, and think them of little use in the diet.

Most puppies are born with round worms, and must be treated for such. Six weeks of age is the usual time to do this, but if the puppies appear thin, and 'pot bellied' after feeding, the coats dull and staring, this may be done as early as four weeks of age. Puppies infested with worms never thrive. There are many very good worming medicines on the market, but if the puppies have to be treated as early as four weeks old it is advisable to ask your veterinary surgeon for some pills. The amount to be given should be judged by the weight of the puppies, and the dose varied accordingly. The treatment should be repeated in eight to ten days' time.

At seven to eight weeks old each puppy should be having halibut or cod-liver oil: one drop each day for the first week, gradually increasing the dose as they grow, until they are each having about half a teaspoonful of cod-liver oil a day, or three drops of halibut oil.

Personally, I find the four to eight weeks of age the most difficult of all with puppies. Fox Terriers are a hardy breed, and things seldom go wrong while weaning them, but it is a time when one must be watchful and attentive to their requirements. They tire very quickly when so young, and need very little exercise. They play with each other for a short time and quickly fall asleep, all curled up together in a warm bed.

Here I feel I must stress again the absolute necessity of a warm, dry bed. Warmth, to my mind, is every bit as necessary as good food.

Bringing up a litter of puppies is hard, exacting work, but most interesting. No two puppies are alike, and each different character can be seen in every litter at a very early age. Perhaps there is a bully, perhaps a timid one, even a cunning one. Puppies are hard task-masters, but, to me, worth every minute wasted on them. Yes, they definitely are time-wasters. Over the many years I have been breeding Fox Terriers I have wasted hours and hours watching puppies grow and develop into smart young dogs, and I have thoroughly enjoyed every minute of it.

When the puppies have reached the age of eight weeks that is the best time to dispose of any which are superfluous to requirements. They are most attractive at this age, and many people like to have them young and to bring them up themselves. Purchasers should be strongly advised to have their puppy inoculated against distemper and hardpad. My veterinary surgeon advises that this should be done when the puppies are nine to ten weeks old. I have found that 'Epivax' injected at this age is 99 per cent safe against these foul diseases. Remembering how difficult it used to be in the good old days to rear puppies successfully to the age of six months or a year without the help of these modern drugs, I must stress the advisability—no! the necessity—of all puppies being inoculated as soon as they are old enough. I have seen, during the years when the drugs were not available, whole kennels of youngsters wiped out with these heart-breaking and distressing diseases.

If any of the puppies are sold as pets before being inoculated the purchasers should be told not to take the puppy out before he has been immunized. The danger is not so great these days, because most dogs, I think, and hope, have been made safe with inoculations to mix with their fellows.

Purchasers of puppies should always be given a feeding chart, so that a complete change of food may not upset the puppy's stomach. I advise such people to feed them as they would a young child, giving the puppy a little of anything children have, such as milk puddings, custard, fish and meat with brown bread, but never potatoes. Of course, in a Kennel this cannot be done, because there would not be enough scraps from the table of an ordinary-sized family to go round.

The four meals and the hot drink at night should be continued until the puppies are three months of age and then reduced to three meals a day. Breakfast: corn-flakes and milk, cod-liver oil and a calcium tablet each. Dinner: soaked biscuit meal and minced meat. Supper: similar to the dinner.

At about four months of age the puppies start teething, and this goes on until they are eight months old. Two meals a day should now be sufficient, but much depends on the weather and the individual terrier. Some put on weight and condition much quicker than others, and the racy types seem to require much more food. In cold weather they need ample food to keep up the body temperature.

It is a good plan to give all puppies a teaspoonful of milk of magnesia one day a week while they are teething, and large raw marrow bones to chew. Allow them to play with a sack or an old blanket, so that they can pull at it one against the other and so pull out any loose milk teeth.

When Fox Terrier puppies are teething they are inclined to carry their ears erratically. Some people advocate sticking them on to the head in the correct position with sticking plaster or adhesive liquid. I find it much better to leave them alone. Correctly placed ears invariably settle down in the correct fold as soon as the teething is finished. It is, however, not advisable to strip ears during teething; it is better to leave the hair, both on the inside and outside. Taking all the hair off at this time often makes the ears 'fly'. The owner then, in desperation, either sticks them down or continually rubs oil on them, which to my mind makes the terrier conscious of his painful ears. The sticking plaster makes the ears sore and the muscles behind the ears stronger, causing them to fly even more.

Fox Terriers are real terriers and rarely more than two can be housed together. Some litters start fighting at a very early age, and simply have to be separated at about eight weeks old. I make it a rule, once they have started fighting among themselves, never to leave three together, as two against one is not fair, and may be dangerous. Two in one kennel is all right. If they do have a difference of opinion it is a fair fight and usually over very quickly.

I find, too, that if one of the puppies is taken from the others for any length of time, for one reason or another, upon returning, they all pitch into it at once, sometimes inflicting very serious injuries. I suppose this is caused through jealousy.

It is noticeable, too, that if one of the puppies is at all ill, the stronger ones turn on it unmercifully. This, presumably, is the law of the jungle—the survival of the fittest.

Rearing puppies is definitely a difficult and exacting task. One must always be on the watch for any new mischief they can be up to. It is surprising the things they can find to do—digging holes, chewing kennels and anything else they can get their teeth into.

Amusement should be provided for them in the shape of a hard rubber ball, an old leather shoe or, as I have said before, an old blanket or sack to play with in the runs.

Good food, clean water, and a cosy bed are absolute essentials. Plenty of sleep for growing stock is most important.

The reward of all this care is the growth of one's puppies into sound, healthy stock, a credit to the breeder, and a source of satisfaction and pleasure to the owner or purchaser. Should one be so skilful as to have produced a winner, then the triumph is complete.

Before the puppies are ready to leave the mother at eight weeks of age the prudent breeder will have to decide their future—are they all to be kept or are some to be sold? The latter is the most likely course.

While the puppies are running about in a run or playpen they should be watched and studied very carefully. Even at this early age it is quite easy to see which have the most 'style'. Style can best be described as smartness plus personality. A Fox Terrier is a smart, tailor-made dog when grown, and some puppies show this characteristic far more than others. Character, I feel, plays a very large and important part towards style. The puppy that 'pulls itself up' and stands to attention, thereby showing his good points naturally, is certainly one to consider keeping from the litter.

Some people have an 'eye' for dogs—or for any animal—and in selecting puppies at an early age it is a wonderful gift to possess.

If at all possible it is advisable for beginners to ask an acknowledged expert of the breed to give the benefit of his advice. I am sure that most fanciers are interested in any well-bred litters and very willing to help in deciding the best puppies for the owner to keep.

Quite often I find that when I have been invited to look at other people's litters the owners have already decided which they are keeping, not necessarily from a show point of view, but usually because it is the prettiest in colour, or the one with the most amusing character. Sometimes the smallest one, simply because they feel sorry for it.

Time spent watching the puppies in their playpen is time well spent, and to my mind the only way to decide which to keep.

Personally, I do not like an even litter; that is, one in which all the puppies are very similar in make and shape, and much the same size. I find it much easier to come to a decision when there is perhaps one completely outstanding in show points, such as a long, lean head, reachy neck and sound quarters. In an uneven litter there may be a more cobby type of puppy, with a very short back, and lots of substance, but naturally not so long or lean a head. Then, too, there might be a smaller one altogether, cute, very pretty and attractive.

When sorting out puppies to keep as potential show stock, having decided which are the natural 'showers', stand them one by one on a table and look at them well.

In my opinion dog puppies must be really outstanding if they are to be kept. There is an old saying that dog puppies are worth £5 or £50 in the nest. This is not quite correct at the present time because strong, healthy pet puppies can readily be sold for fifteen guineas or twenty guineas each at eight weeks of age, and a champion-bred, very promising dog puppy is worth far more than £50.

My advice, especially for the beginner, is to part with the dog puppies at an early age if they have even the slightest fault, and it is rare indeed for this not to be the case. It is far easier to sell dog puppies young, because if a dog puppy has been brought up in a kennel until about six months of age it will have lost its pretty puppy character and appearance. The pet buyer will therefore be less attracted by it, and the buyer requiring a show specimen will naturally be more critical.

When examining bitch puppies, if any one of them has an uneven set of teeth, whether undershot or overshot (see Fig. 2), my advice is, however good she otherwise be, not to keep her. I find that when the puppy's second teeth come through they rarely come straight. An extraordinary thing I have noticed also is that most puppies with either undershot or overshot mouths have 'apple heads'. This simply means that the skull is rounded and bumpy.

Many fanciers have been heard to say that the bump will flatten in time and grow down into the foreface. I have found this to be wishful thinking; the heads do not flatten nor do the teeth become level.

The standard clearly states that a mouth undershot or overshot is a disqualifying point, so should you find one with this bad fault it is advisable to sell her as a pet.

Ears, too, must be carefully considered, as they also are mentioned in the standard as disqualifying points if not correctly carried. At the age of about eight weeks I prefer them to be on the large side, set high on the head, not hanging at the side of the head like a hound's, but large enough to fold correctly, and to lift a little while the puppy is teething. Not all small ears 'fly' permanently during and after teething, but there is a danger of their doing so. Therefore, keep a puppy with the larger ears.

Eyes are more difficult to decide about at this age, as they invariably have a blue look about them. Any, however, with large, round eyes would be discarded.

Soft coats are an abomination, and although they can be improved by constant stripping, they are always obvious to a trained eye, and a continual worry to a would-be exhibitor. Even if one is able, by repeated stripping, eventually to grow a fairly good coat it is not satisfactory because the coat only stays tidy for a very short time, and the dog seldom looks in good show form. A well-coated dog can be trimmed over and over again, and can look well enough to be shown for as long as a whole year. It is sometimes difficult to decide whether the coat will eventually be soft or not. The best place to feel for the texture is either on the rump or on the tail. Soft coats are a definite defect, so choose a puppy with the promise of a hard coat, remembering that the coat improves slightly with the first stripping or two.

It is practically impossible to decide the size that each individual puppy will be when fully grown. So much depends on their upbringing, environment and breeding. The thick-set cobby type does not seem to grow as large as its more elegant racy-type brothers and sisters. Puppies with large, bony knee-joints usually grow into big adults. I do not advise selling puppies purely on consideration of size because anything can happen about this. A slight illness may cause a set-back, and lack of warmth, or incorrect feeding, can easily retard growth.

Colour, to me, does not matter a bit. A plain, almost ugly, white puppy, with poise and outline, I think, is far more attractive than a beautifully marked one that does not 'swank' and is sloppy-looking.

So when selecting a bitch puppy for show and breeding later on choose the one with the lean head and flat skull, larger ears, long neck, good shoulders, level back, with good set-on tail, and strong, sound hind quarters.

Neck and shoulders, and therefore the poise of the head, greatly add to the quality of an animal. Should it have to be a choice between a puppy with a long neck, good shoulders and a little length of back, and a puppy possessing a very short back but with a short neck and upright shoulders, I would definitely choose the former.

Fronts, coats and feet may improve, but the puppy with style and quality which is so attractive is always obvious from a very early age. In my experience these characteristics never develop later if they are not evident in the puppy when young. It is therefore wise to keep the puppy which is promising in this respect.

When choosing a puppy as a pet it is well to remember that it must be strong mentally as well as physically. Pet puppies do not live such a sheltered life as kennel dogs. They are often taken here, there and everywhere, at a very early age, and must find the noise and bustle of life most bewildering. Therefore, choose one that comes forward when you go near the kennel. Never take one that runs away and hides at the approach of strangers. Shining eyes, wagging tail and a firm, strong body, are things to look for. If the one really preferred proves to be a little more expensive than the others, remember that the good-looking pet will provide pride and pleasure for many years. It will cost no more than a new coat, and it will certainly last longer.

The choice and purchase of a puppy, whether for a pet or for show purposes, is an exciting experience, and worthy of considerable care and trouble. It is wise to visit several Kennels in search of the right one. The result of a correct selection is many years of satisfaction and happiness.

CHAPTER 8

STRIPPING AND TRIMMING

ALL owners of Fox Terriers have in their mind's eye a picture of their dogs as they would wish them to appear: that is, smart and attractive and similar to the terriers they see exhibited successfully both by amateurs and professionals at the local and national shows.

Many people who are inexperienced with terriers do not realize the great deal of work which has gone into preparing the dogs for show, and, indeed, maintaining the animals both in health and coat to look their best.

Smooth-haired Fox Terriers have a close, short coat and therefore do not require the constant attention that is necessary for the Wire-haired Fox Terriers, if the latter are not to appear like unclipped poodles or woolly lambs.

The coat of a Wire-haired Fox Terrier should be, as the name implies, crisp to the touch and, with the aid of the soft undercoat, weather-resisting. It should have the texture of a close, hard door-mat. Unfortunately this is rarely achieved naturally, since if nature is allowed to take its course the coat will become long and open. Periodically, if left unstripped, moulting would occur and the owner would soon complain about the cast hairs. The dog would not only look most untidy but would also be very uncomfortable.

To overcome this characteristic of the breed, the coat therefore has to receive periodical attention. If the purpose in owning a Wire Fox Terrier is to keep him as a pet only, then the work entails a stripping about twice a year. If, however, the idea is to show the dog, then, in addition to stripping, there will be a considerable amount of trimming before he can be exhibited with any hope of success.

When the puppy has reached the age of about eight to twelve months he or she will be in 'full coat'. This means that the outer hair or the top coat has grown to the length of two inches or so on the head and body, his whiskers and leg hair being even longer. Also, as the Wire-haired Fox Terrier has a warm, soft undercoat, this will have, in all probability, grown thick and woolly. In consequence, the leg hair, and particularly under the armpits, tummy and the underneath side of the hind legs, will quite likely have become matted and in tangled knots.

The Wire-haired Fox Terrier should, and must, be a smart dog in appearance, with his crisp wire coat fitting him like a glove.

In order to grow such a coat the first thing one must do is to 'strip' all the long, old hair off.

This is done with a stripping knife, which can be bought quite cheaply at any pet stores, or at the trade stands at any of the big dog shows. In choosing stripping and trimming implements they should be held in the hand before buying, to find out which feels the best. They will be used for a great many hours, and to have comfortable-feeling implements is most important, because even when using tools easy to hold and comfortable in the hands they can still cause corns to grow on the hands. It is best to stand while stripping, with the dog standing on a table of a height convenient for one to 'stand over' the work.

Stripping is done with a stripping knife—one with serrated edges is the best as it gives a better grip of the hair. The knife should be held in between the first finger and thumb, and across the palm of the hand, the long hairs being pulled out at a few at a time in a quick, sharp movement, pulling towards one's own body. There is a right way and a wrong way to strip, as in everything. Never pull against the growth of the hair, as this will make the skin pink and sore. It is particularly important to pull the hair the way it grows. Always hold the skin tight with the free hand before pulling, and only pull out the hard, outer coat, leaving the soft undercoat for the time being.

Starting with the head, with the dog either sitting or standing on the table or bench, strip from between the eyes, over the back of the head. Do not take all the hair off over the eyes, but leave about an inch of hair to form eyebrows, which can be shaped when trimming for show. This greatly improves the terrier's expression. Tidy the hair on top of the muzzle into a straight line from the eyes to the nose, but be careful not to take too much off here, or the face will look pinched. The muzzle and jaw should appear strong. Never strip muzzle, legs or feet, but rather trim them. The ideal effect is for the

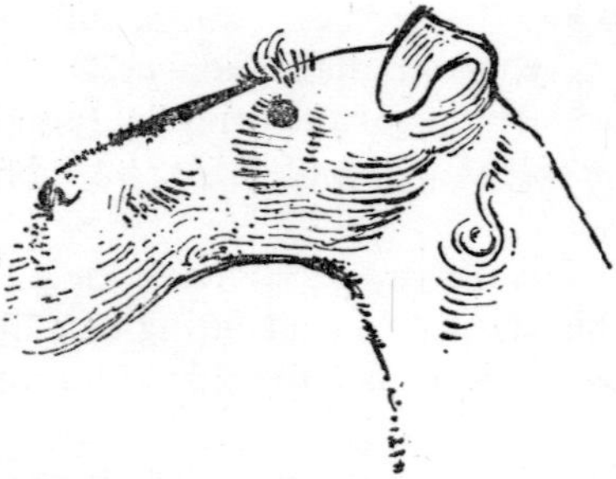

Fig. 12 Crown of neck hair

head to appear about the same width across the skull as across the muzzle. This not really so, as would soon be seen if the dog got his muzzle wet. Under the hair the head will be found to be well chiselled, but present-day fashion demands plenty of whiskers and a long, lean head. The whiskers help this illusion. That is why it is advisable not to comb the muzzle hair unless it is absolutely necessary to remove knots or particles of food, but preferably to brush or rather 'ease' it out with a wire rubber-backed pad. Whiskers and leg hair, once removed, take many months to grow.

Next, strip the cheeks close to the head, and continue down the side and back of the neck. It will be noticed here that the hair grows into a crown on either side of the neck, so pull the way the hair grows (Fig. 12).

Now we come to the ears. These are most important, as they affect the whole expression; they should be kept very short of hair. Pull the long hairs from the outside of the ears, being very careful not to allow the serrated edges of the stripping knife to dig into the skin, as this causes sores to form, which are very difficut to heal, and often make the dog carry his ears badly. I use an old penknife for this. Having done the outside of the ear, turn the ear over and very carefully do all round the edges of the underside, finally trimming out all the long hair from the inside base. When the ears are finished, and look neat and tidy, smear olive oil or glycerine over the outsides. The ears are very sensitive, and oiling prevents the terrier becoming 'ear conscious' and consequently flying his ears. By 'flying' I mean the ears are carried away from the head in a high position.

Having finished the head and ears, put a collar and lead on the dog and fasten him to a hook in the ceiling over the table or bench. This keeps the dog in a standing position to enable one to continue more easily. Now strip down the shoulder blades and down the sides of these to the top of the forelegs. Next, with the left hand, hold the head up and strip under the chin, starting about two inches from the end of the muzzle and pulling all the hair out from around the throat. Here will be found two more crowns and it is best to strip upwards from the front of the forelegs (Fig. 13).

Do not strip the forelegs. They have to be trimmed later on to look

Fig. 13 Crown of chest hair

absolutely straight viewed from any direction. Any long, straggly hair may be pulled out, great care being taken not to take off too much, because leg hair takes so much longer to grow than the body coat, and should always be much longer, to comply with present-day fashion. Before taking any leg hair off, comb the hair through with a wide-toothed steel comb straight down the front and backwards from the elbow. Pull out any long hairs with finger and thumb from the front of the legs, and straighten the appearance from the sides and back of the legs by carefully pulling the long hairs from the fringe which runs down the back of these limbs.

The toenails will need attention. If the nails have been allowed to grow long—due in many cases to lack of exercise on hard surfaces, or through running on carpets or other soft surfaces—they can be cut with nail-clippers. However, filing is better, because if once the nails have been cut too close to the quick, thus causing pain, any sensible dog will remember, and the next time anyone tries to do anything at all to the nails there will almost certainly be a battle of wills. Sometimes, if a dog so much as sees a file or a pair of nail-clippers, he prepares for a battle. Should this happen, the next time you are preparing to file the nails, stand the dog on the table and put the file near his feet. Do not attempt to use it, but brush the dog and so show him that the implement will not hurt him. Gradually he will allow you to touch his feet with the file, and within a few days he will let you file his nails, just a little bit at first. Great patience is required to make a dog forget a hurt, and prevention is better than cure. That is why I always use a file in preference to clippers.

To file the nails, hold the leg firmly in the left hand, grip the dog tightly under the left arm and then file each nail to the quick, which can easily be identified by a pinky colour in the light coloured nails. Black nails must be more carefully filed, as it is not possible to see the quick. Use a flat rasp (file) about one inch in width, making strong firm strokes, and filing sideways, finishing off with smaller strokes to make the nail smooth and clean (see Fig. 14).

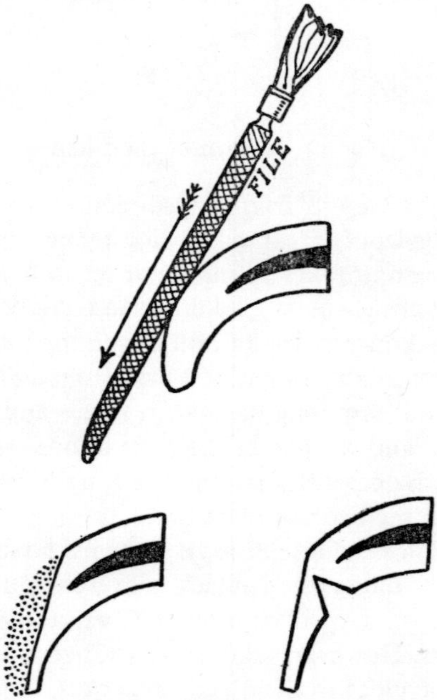

Fig. 14 Filing the nails

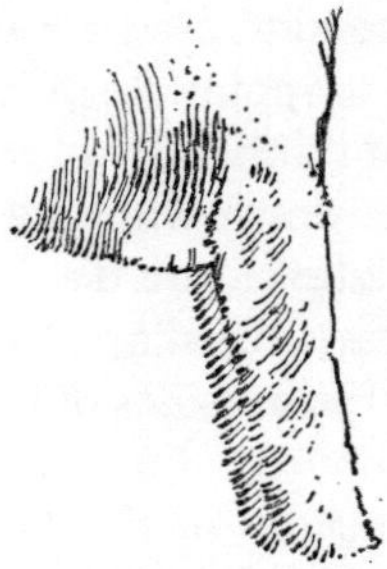

Fig. 15 Front legs correctly trimmed

Still holding the leg in one hand, cut the hair from underneath the foot in between the pads with a sharp pair of scissors, and then trim round the front and back of the foot, either with the stripping comb or a penknife.

When the feet and nails are finished they should appear as round as possible, giving the impression that the terrier is standing on its toes (Fig. 15).

Fig. 16 Crowns of hind quarters

The next thing to do is to start stripping from the back of the head (the occiput bone), right down the top of the neck, over the withers or point of the shoulder blade, along the back to the root of the tail. Then, strip both sides of the body, over the ribs, loins and hips, leaving the remainder of the hind legs until later.

Next, take all the long hair off the tail, pulling the hair upwards from the backline, both in front and at the back of the tail.

There should now be only the hind legs and hind quarters to do.

The hair on the back of the hind quarters grows in two crowns. The terrier is very sensitive to stripping on this part of the body, so do this particularly carefully, taking only a few hairs at a time, and not hurrying. He will try to sit down while you are doing this, so hold him up by the tail, and whatever you do, do not lose patience. Strip up from the base of the heart-like shape formed here by the way the hair grows, and pull down the outsides of the quarters, not forgetting to continue to pull only the way the hair grows (Fig. 16). Take all the long hairs from the outsides of the hind legs and trim the hind feet in the same manner as the front feet.

The dog should now look tidy, though a little naked, because he has only his underclothes on.

Trimming and maintenance of coat

After stripping, the outer hard coat takes about eight weeks to grow through the undercoat that is left on, and during this time it is most important to use an undercoat comb to prevent the undercoat from growing at all long and thus preventing the new hair growing. Use the special comb made for the purpose, which has a blunt blade as well as the serrated edge. This, again, may be bought from any dog shop. Simply comb the dog all over—except his legs or muzzle—holding the comb flat to the skin, so that the sharp edges of the comb cannot scratch the skin. Doing this for a short time every day, or on alternate days, it is surprising how much fluff comes off. It is like weeding a garden, one must take the weeds out to give the plants a chance to grow.

Trimming for show is quite an art, and should result in making the terrier look his very best. In preparing him for show when his crisp, new coat has grown, stand him on the stripping table and look at him closely from all angles. It is a good idea to stand him in front of a large mirror, so that the way he looks from the other side, as well as from the front and back, can be seen. To my way of thinking, trimming is the practice of hiding the dog's faults and accentuating the good points. That is why I say, 'Look and look again.' Having decided where it is necessary to alter him, set to work slowly and carefully.

Remember, you have quite a long time in which to prepare your dog for the show, and the poor judge has only a very short time in which to find the faults you so carefully hide! They all have their faults you know—the dogs I mean, not the judges! It is necessary to

Thurse

Ch. Cawthorne Climax (Born 1953)

Ch. Cawthorne Chloë (Born 1954)

Thurse

H. J. Goater

Ch. Gosmore Rosemorder Fireaway (Born 1952)

Ch. Penda Callern Melody (Born 1954)

Thurse

start about three weeks before the actual show, trimming to bring the dog to perfection on the great day.

Should, for instance, his head be a bit coarse—that is, wide through the skull, or with prominent cheeks—take the hair off as closely as possible. This will give the appearance of the long, lean head, so much desired (Fig. 17).

Should there be a dip behind the shoulders (the withers), in appearance this can be improved by allowing the hair to grow in the 'dip' (often described as having a dip in back), making the back, to all intents and purposes, strong and straight: in other words, giving him a good top-line (Fig. 18).

Then again, should he turn his hocks in (known as cow hocks), strip the hair off the inside of the hind legs to make them appear straight (Fig. 19). You will notice I say 'appear straight'. This is

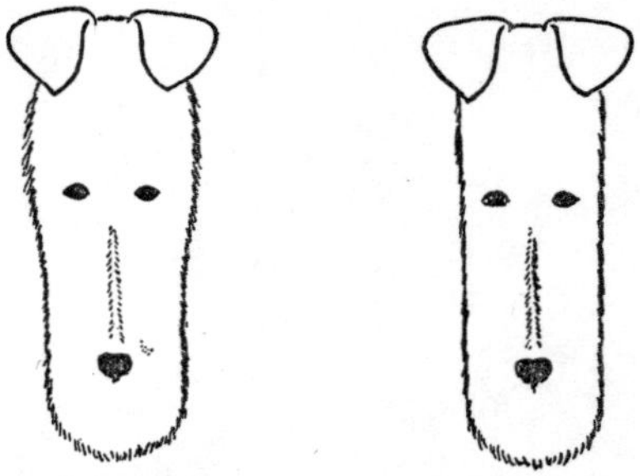

Too wide in skull. Skull hair trimmed.

Fig. 17 Trimming of head

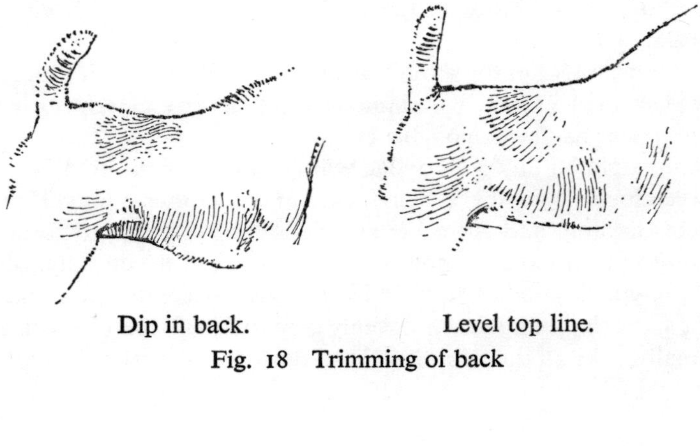

Dip in back. Level top line.

Fig. 18 Trimming of back

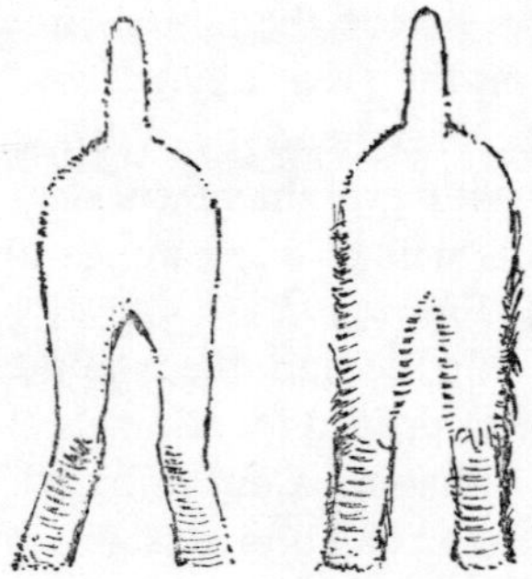

Cow-hocked. Trimmed correctly.

Fig. 19 Trimming of hindlegs.

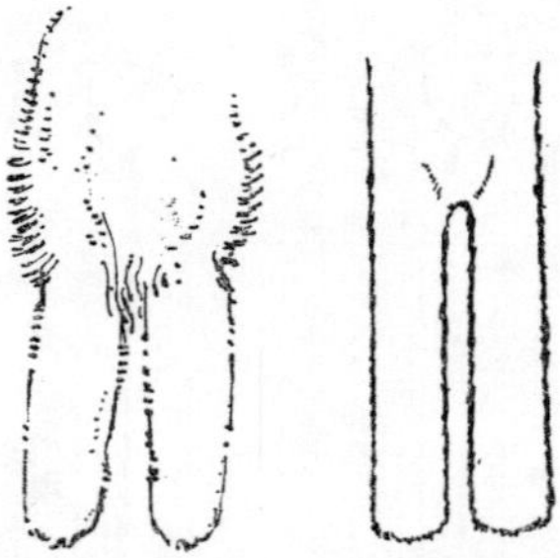

Loaded. Fined down.

Fig. 20 Trimming of shoulders

intentional, as no amount of trimming will make good quarters out of bad ones—it is just a matter of doing the best one can with the material available.

The appearance of the shoulders can also be improved, if they look loaded from the front, by trimming the hair very closely to make them appear as flat as possible (Fig. 20).

In my opinion any good judge will quickly discern the difference between looking good and being good in the various points. He will handle the dog, and from his experience will be able to form an opinion as to whether or not the animal is sound structurally. But a well-presented, good-coated terrier will always command attention in the show ring, as against a slovenly prepared, soft-coated specimen.

Finally, take all the hair off tight to the belly, and trim the hair on the brisket and loin with a pair of scissors, to make as good an outline as possible.

When all this has been done you will be able to take your dog to the show in the knowledge that you have done your best to give him the greatest chance of winning his classes.

An untouched photograph of Ch. Penda Peach, owned, bred and handled by the author, which illustrates the result of careful preparation as I have described, is reproduced among the pictures.

The Smooth variety of Fox Terrier does not need stripping, but every day a little attention should be given to the coat. Brushing with a hard brush helps the new coat through. An undercoat comb helps to remove any dead hair. Do not pull the hair, but simply comb through the coat. Ears, cheeks, throat and neck can be improved in appearance by slight trimming, also the back of the tail and the frills on the back of the trousers should be removed. The feet and nails should also receive attention similar to the Wires.

The great thing to remember is that the Smooth looks very bad if the coat is cut in any way. It is very difficult to trim a Smooth evenly without its being obvious. It is considered better to show the Smooth variety of Fox Terrier untrimmed, rather than 'cut about'.

The first time I tried to strip a dog I became most depressed and disheartened. I thought I would never be able to do it correctly and master the art—and art it is.

Always remember that no two people trim alike, and no two people give the same advice. Perseverance and patience are absolutely essential. Use your eyes, and watch the professionals. These are a group of men who prepare and handle dogs for show. Quite often an owner has not the knowledge or the desire to show his dog himself. To make a dog a champion takes a tremendous amount of hard work, both in preparing him for show and in travelling to the various shows. To become a champion the dog has to be adjudged the best of its sex at three different championship shows under three different judges. To do this they must be in tip-top condition every time they appear in public, and the professionals are certainly able to ensure this. It is their chosen job in life and they keep the competition hot and the standard very high. They are true sportsmen, who know how to win well—and lose well. They are kindly, clever men, and when the judging is over, and they have finished their work for the day, they will do all they can to help.

CHAPTER 9

PREPARING FOR SHOW

PREPARING a terrier for show entails a great deal of patience and hard work. Patience, because, as well as the stripping and trimming, the dog to be shown must be fit both mentally and physically. This requires much time. He should be shown in good condition, which means neither too thin nor too fat. Common sense must prevail as to whether to give more food or less before the actual show date. No two dogs are alike in their likes and dislikes and they therefore must be studied individually and fed accordingly. To me, a fat terrier is even worse than a thin one. When I say fat I mean really fat, so that the shape of the body cannot be seen. When looking over the top of the back the spring of ribs and the loin should be easily discernible, and so make a good outline. The body should not be like a great fat sausage, nor, for that matter, like a drainpipe.

Having previously stripped your dog, as described in the previous chapter, your terrier has to be educated, which means that he must be taught to stand and to walk correctly on a lead. This should start not later than about eight weeks before the show at which he is to make his début. He should have a lesson on the lead each day, starting with short lessons at first, just once or twice up and down in a straight line is the best way to begin. When he has learned to walk properly, neither pulling, nor jumping, he must next be taught to stand still and to show himself, with his head and tail high.

When teaching the dog to walk properly do not allow him to pull on the lead, because a dog that is pulling never moves correctly but 'straddles' with his hind legs, and tends to move sideways instead of in a straight line. With practice, one can tell by the feel of the lead whether the dog is going well or not. It is very similar to handling the reins when driving a horse on a tight rein. If the terrier is checked when he pulls he will soon move on a loose lead, which is most desirable. Should the dog pull persistently, it is a good idea to roll up a newspaper and hit him across the nose with it. This does not hurt him at all, but it makes a frightening noise and teaches him not to pull the next time you ask him to walk.

He must also be taught to stand correctly on a table, because most judges prefer to have the exhibits standing on a table to enable them to be examined properly. Therefore, when the dog has been taken up and down several times, lift him up and make him stand still on a table. Do be careful to have a firm table handy, as all dogs hate a wobbly table and they feel most insecure when placed on anything that is not firm. A little and often is the secret of success in these very important exercises.

When the dog has become accustomed to the lead and has confidence in his handler (or trainer) he must be further educated by taking him out into the street. Do not forget that if your dog has been brought up in a Kennel, and has therefore seen nothing or very little of the outside world, such things as perambulators, bicycles and motor-cars must be very terrifying. They must seem to be great moving monsters to him. If he appears really frightened when he meets them for the first time it is a good idea to pick him up and carry him a little way under your arm. Dogs have great faith in their owners and feel much safer when picked up. So, if he is frightened, take him home and start again the next day. It is absolutely no use at all dragging him along on a lead when he is really frightened. Each time he meets strange things he will grow accustomed to them and, eventually, take no notice at all. Most Wire Fox Terriers are terriers, and usually fearless. Therefore, they quickly learn just what is required of them and soon become completely manageable in every way.

The great thing to remember is that, when taken from the Kennels at first, they, like young children, are uneducated. Time and 'homework' alone will remedy this.

As an instance of the way familiarity can overcome canine fear or nervousness, I can recall the experience with a Foxhound bitch, Avon Vale Vampire, which was given to my husband some years ago by the then master of the Avon Vale Hounds, Mr Holland-Hibbert (now Lord Knutsford). My husband was returning to Trinidad, British West Indies, from leave and wished to take out a Foxhound or Beagle puppy in order to hunt the native deer, as the local dogs had little power of scent and often lost the line. On enquiry at the Avon Vale Kennels he was informed that puppies were not for disposal, but he was offered a four-year-old bitch, Vampire, whom the master intended to put down owing to her habit of riot. This bitch had hunted three seasons with the pack after returning from walk on a farm. She therefore appeared ideal for my husband's purpose and was gladly accepted.

On being taken from the Kennels, Vampire immediately showed great fear and nervousness when confronted with road traffic. This was something quite outside her experience and was terrifying to her. The cure therefore was, little by little, to increase her familiarity with traffic. In time this treatment was completely successful. Not only did Vampire become entirely indifferent to the rush and noise of traffic, but she also even came to like her travels on car or lorry. Her final achievement was to ride between the mudguard and bonnet of one of the company's lorries, to which position she immediately sprang as soon as the engine was started. There she would ride, pressing her body against the bonnet to prevent herself from falling off.

While teaching the dog to stand properly, I find that having a brush in one hand is most helpful. Rubbing the brush gently over the back will make him hold his tail high, and brushing with a downward movement over the sides of the shoulder blades will teach him to stand with his front feet together. Do not place your dog in the correct position by picking him up by his tail; he will hate it. By far the better way is to handle him under the quarters (in between the hind legs) and gently lift him into a good position, with his hind legs well out behind him.

When he is standing still, pat his leg hair into place, and also brush his face hair (the beard) forward. These little bits of brushing help to keep the dog still, and with the aid of the brush he can be made to look his best while standing. Dogs like being fussed in this way and soon understand when you are pleased with them.

After he has grown accustomed to walking and standing still correctly, and is no longer afraid of the everyday things he sees near the home, take him out in the car for a short ride and then walk him home. The next day try to take him for a little longer ride and so on, until he looks forward to his car journeys. 'Pub-crawling' is marvellous for his education. Most people, even if they do not like dogs, will usually take kindly towards your dog when they have had one or two drinks and will come and pat him. This is worth untold wealth, because it teaches the dog that strangers are his friends—not his enemies.

I find that going into hotels to train my puppies is one of the more enjoyable parts of preparing for show, for not only does it encourage the puppy, but it also encourages me, and makes me feel very hopeful for the next show I have in mind.

It is quite obvious that dogs experience enjoyment when their owners are feeling cheerful. Just as he knows when his owner is feeling depressed, so he also knows and appreciates when his owner is happy. Some people firmly believe that one's feelings run down the lead. This, I believe, is an old wives' tale, although dogs certainly understand intonation of the voice. To prove this point, try saying such words as 'good dog' in a pleasant voice and saying the same words in a harsh, hard way, and notice the difference in the reaction. I find that one can even grumble in a nice way and dogs do not mind at all, but try really grumbling, with an occasional jerk on the lead, and see what happens! Of course, some dogs are much more sensitive than others, and care must be taken not to upset a youngster.

That is why I advise patience, and again more patience.

While this preparation of the mind is going on, the coat must also be kept in good condition. This is done by brushing daily until the wiry outer coat comes through and is the desired length. Once the outer coat has come through, an undercoat comb should also be used daily to keep the undercoat short and so allow the outer coat to lie flat and really fit the body.

When your puppy is standing on the table it is a good idea to open his mouth occasionally, because when he is being examined on the table by the judge his mouth will have to be opened either by you or the judge. This is for the judge to be able to examine the teeth for any irregularities. This is all part of the puppy's education and he will soon become quite accustomed to a complete stranger opening his mouth.

Also make sure the teeth are clean. If you are showing a young dog they most certainly will be, but sometimes an older dog's teeth become coated with tartar, which is easily removed by scraping the teeth with the serrated edge of a sixpence. A sixpence is ideal for this job, since there then is no danger of injury to the gums as there would be from a sharper instrument, should the dog suddenly move.

When the puppy or young dog has had his daily exercise and grooming, now is the time to teach him to rest contentedly in a travelling box. The box should have a clean piece of blanket in it, preferably a white blanket, as colours sometimes come off and stain the coat. White ones ensure that the white coat of the dog remains clean. Nothing appears worse to me than a white-coated dog looking grey.

Talking about using coloured blankets reminds me that many years ago I went to inspect a litter of puppies which were about eight weeks old and owned by a complete novice. The puppies were in an open basket, asleep on a bright pink blanket. They looked fat and well, but the owner seemed most disappointed when I told him that they would never be show specimens. He asked my reason for saying so, and I pointed out that they all had liver-coloured noses! It took me quite a long time to persuade him that the noses really were liver. He told me, in all seriousness, that it was only the colour which came off the blanket. I am pleased to be able to say that I have not seen puppies with liver noses for a great many years.

The dog should be left in the box for a very short time at first. It is a good plan to give him a sweet biscuit, or any titbit of which he is particularly fond, every time he is put into the box. I have had dogs on occasion that, on first being introduced to their travelling boxes, really resented the treatment. But a titbit works wonders, and soon you will find that your dog will run into the box as soon as you touch the biscuit tin. All my show dogs do!

It should not be necessary for me to remind you to manicure the toenails, if required, as described in a previous chapter. This treatment must be continued right up to the day before the show.

Continue each day, if only for a few minutes, the exercises on the lead and the table, until you are quite satisfied that the dog will look and behave like the aristocrat he is, under all circumstances correctly, and in any company.

The next thing to practise is the actual preparation of the dog, which has to be done prior to his going into the show ring. Most exhibitors have their own way of doing this, and no two people do it in exactly the same way. As many people have asked me how I prepare mine, I will describe just what I do. I do not say mine is the

Thurse

Ch. Penda Oregan Witchcraft (Born 1957)

Ch. Penda Peerless (Born 1959)

Thurse

Thurse

Ch. Hermon Rebel (Born 1956)

Ch. Hermon Fantasy (Born 1960)

Thurse

best or even the correct method, but the final result is quite good, I am told. This preparation, the dog thoroughly enjoys, provided you handle him firmly, but kindly.

First, wash the leg hair and face hair (the beard) in warm, soapy water (not detergents), and dry gently with a soft towel. I say gently purposely, because if the hair is rubbed and dried too roughly, or with too harsh a towel, it tends to break off, or to fall out. The fashion for the show ring today is for plenty of short, strong leg and face hair which can be brushed into place, rather than the long, loose sort of leg hair which, if allowed to grow too long, is usually very soft and undesirable.

Then rub a little hair cream, such as Brylcreem or white Vaseline, into the face and leg hair, and with a soft bristle brush dab in plenty of powdered chalk. Ordinary white chalk will do, but precipitated chalk is much better. I, myself, use Robin starch because I find it sticks to the hair very well, and it certainly gives a much whiter finish than any of the others. Never rub the legs with knob whitening. It is much too hard, and takes more hair off than can be grown in a month of Sundays. When the leg hair and whiskers have been chalked they will stick out in all directions, but leave them as they are for the time being. Next, with a piece of ordinary knob whitening, rub well all over the coat, stroking the coat the way the hair grows, i.e. straight down from behind the ears, down the neck and straight down the whole of the back and body towards the tail. Allow the dog to shake himself to get rid of the superfluous chalk and then brush him all over, with the exception of the face and leg hair, with a soft brush.

The leg and face hair have to be patted into place with a wire pad. This is a special rubber-backed brush with wire bristles, rather like an old-fashioned curry comb. Do not comb the leg hair, because once it has been pulled out with combing it takes a very long time to grow to the correct length again.

If the dog is hound-marked—that is, has a tan head with black body markings—put a little hair oil on a soft brush and brush all the coloured patches. This makes the colours much brighter, and the contrast with the remaining white parts looks most attractive.

The Smooth variety of Fox Terrier requires the same training and education as I have described for his Wire counterpart, but he does not require nearly as much actual preparation before showing. His legs and face only will need washing, and the whole dog should be chalked all over before combing and brushing. Any markings he may have can be brightened up in the same way as for the Wire.

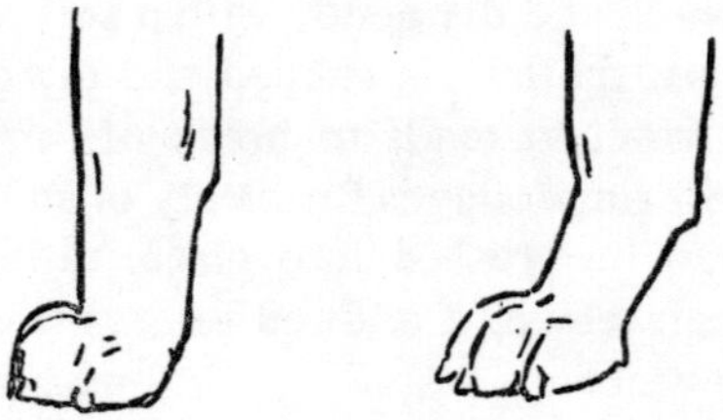

Right. Wrong.

Fig. 21 Feet

His coat may require tidying up from time to time, and this should be completed about fourteen days before the show.

In a previous chapter I have given a more detailed description of trimming and, as I said then, which I feel is worth repeating, in

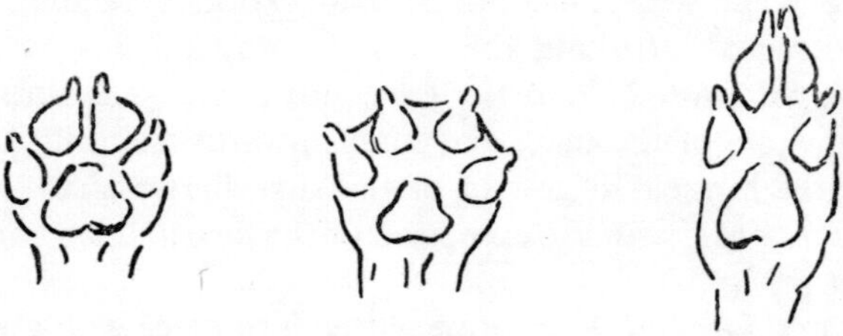

Right. Open. Long.

Fig. 22 Pads

trimming a Smooth for show the great thing to avoid is a cut-about appearance of the coat. That is why it is best to do it a fortnight before the show. Like the Wire, his coat must be kept well groomed right up to the show date, with his coat fitting tight to his body.

Always remember that each terrier requires individual treatment, simply because no two terriers grow their hair in similar periods of time. They certainly do not grow hair the same length all over the body. Experience alone with each dog will teach you the best way to prepare him to make him look his best.

To me, a Smooth Fox Terrier, with his body in good hard condition, his coat smooth and gleaming white, his eyes shining, his beautiful cat-like feet (see Figs. 21 and 22) and his clean outline, makes an attractive picture of quality indeed.

Time spent in training your dog to show off his good points and time spent on his grooming, believe me, is not wasted. The more strange things he is allowed to see, and the more strange dogs he is taught to ignore, the better. If properly trained, when he is taken into the show rings, among as many as perhaps fifteen to twenty dogs, he will treat them with contempt and will show himself fearlessly with a haughty air.

I have known of young enthusiasts entering the show ring with a promising young dog in bad trim and completely uneducated. The dog is quickly noted by a clever, keen observer at the ringside, who offers a comparatively small sum to the disappointed owner, with the result that what eventually turns out to be a valuable dog changes hands far too cheaply.

When this happens it is not always attributable to ignorance on the part of the beginner; sometimes it is pure laziness. The exhibitor who will not take the trouble to prepare his dog for showing deserves to lose.

Any reputable breeder/judge will, I am sure, give his or her opinion on your unshown youngster. Once you have made up your mind that your dog is good enough to win and to keep for your future breeding programme, stick to your opinion and persevere in the face of all opposition. Never fail to give him the best possible chance of success by thorough preparation for show. Patience and perseverance are two essentials in this most fascinating and rewarding pastime. This surely must be a challenge to everyone.

CHAPTER 10

SHOWS AND SHOWING

HAVING bred your first litter or two, and selected the best from these litters, or having purchased a promising youngster, you will, no doubt, wish to start showing. It is important that one should do so, because shows are the shop window of dog breeding. If you are able to show a good type of stock continually, it gains the Kennel a good name and reputation, and there is no difficulty in selling stock from such a Kennel.

There comes a time when the 'show bug' really bites, and once you have been bitten it soon becomes a disease. Once you have started to show and to mix with the other exhibitors, who in the main are great sportsmen, you will find them good friends, particularly to the beginner. Those that are not sportsmen, in the true sense of the word, do not last very long in the show game. They soon tire of being beaten, usually by exhibitors with much better dogs than their own, and quickly fade from the picture. Life and dog showing, to me, is very similar to a game of cricket, and my father (who was a very good cricketer) taught me that the game is never lost until it is won. There is also no doubt at all in my mind in the truth of the old saying, 'You cannot keep a good dog down'. Therefore, provided you show your dog in excellent condition under an authoritative judge of the breed, you will have just as much chance as the next person of firmly placing your foot on the first rung of the ladder of success.

There are many types of shows to which you can take your rod in pickle for his first appearance.

Presumably you will be taking one of the doggy papers each week, either *Our Dogs* or the *Dog World*, or even both. In these will be found the advertisements for the forthcoming shows, giving the secretaries' names and addresses, together with the places in which the shows are to be held, also the time and the dates of the closing of entries.

Before you can enter your dog for a show he must be registered at the Kennel Club. This means that his official name must be recorded at the Kennel Club. Most dogs usually have a Kennel name and a far more elaborate name for the registered name. The registration forms can be obtained from the Secretary of the Kennel Club. Fill in the registration form prior to the closing of entries for the show, and send it to the Kennel Club, together with the appropriate fee, which is printed on the back of the form. In due course, the Kennel Club will send you a registration card with your dog's particulars recorded on it. You may have purchased a previously registered dog, in which case the owner should have given the registration card to you upon completion of the purchase. This means that the dog must be transferred to your name before he may be shown by you. Therefore you should write to the Kennel Club for a transfer form, which must be signed by the previous owner, saying when the transfer of the dog took place. Your dog can then be entered on the entry form for the show in your name.

As indicated, all dogs have to be registered at the Kennel Club before they may be shown, with the exception of exemption shows, as the name implies.

For your first effort it is obviously wise to enter for a local show. Having decided on which show you wish to attend, write to the secretary of the show for a schedule. This will contain the all-important entry form for you to fill in. Entries for the smaller local shows usually close from ten to twenty days before the event is to take place.

It is most important that every detail given in the entry form must be exactly as the printed particulars on the registration form. Any mistake in this, even a simple spelling error, will cause disqualification by the Kennel Club. In this respect the Kennel Club have recently stated:

'We cannot stress too highly the importance of giving the correct registered name of a dog on the entry form for a show. If exhibitors will look at the notes on the entry form they will see that the name of the dog must be identical with its registered name. It has always been the practice of the Shows Regulation Committee to disqualify those dogs which are not entered by their correct registered names, and this disqualification is made, however small the error may be; there are, of course, many degrees of error in such a matter but the

Committee makes its decision strictly on the instruction given on the official entry forms which is that the name of the dog on the entry form must be identical with its registered name.'

There are four different types of shows that you can enter, as well as exemption shows and matches. These I will describe to help you in your decision on which to attend.

Matches and exemption shows

Exemption shows are held by permission of the Kennel Club. For these small shows the dogs need not be registered, and the entry fees are paid on arrival at the show. Four classes only are allowed for pedigree dogs, and there are usually other classes for cross-breds and mongrels. Such other classes may be for 'The dog which wags its tail the best', 'The dog in best condition', 'The dog with the most appealing eyes' and 'The dog the judge would prefer to take home'.

These are the only shows where non-pedigree and pedigree dogs may compete against each other. Exemption shows are usually run for charity, cost very little to attend and make excellent schooling ground for puppies.

Matches are really social occasions run by canine societies. They are most enjoyable, and a great help to a beginner. Each member takes his or her exhibit, of any breed, along and the match is run on a knock-out basis.

Only two dogs are exhibited in the ring at one time, the winner going forward to the next round and so on, until the Best-in-Show, or rather Best-in-Match, is declared by the judge. No prize card or money is given at matches, but the society usually presents a small prize to the winner.

Matches, too, are sometimes run against another canine society, in which case one secretary makes up his team by invitation and challenges a rival club. Competition at these club matches, as can be imagined, is very keen and great fun.

The ladies of the committees usually provide really good home-made refreshments, the atmosphere is friendly and affords the ideal place to complete the training of your future show dog.

Sanction shows

These are the smallest shows held under Kennel Club rules, and do not have more than twenty classes. To be allowed to show at a sanction show one has to become a member of the society promoting the show. The annual membership fee is usually 5*s*., and to my mind is worth every penny of it, simply because it gives the member the entrée to all that particular canine society's activities.

No dog that has won a Challenge Certificate (*see* championship shows) may be shown at a sanction show. Also, dogs that have won more than four first prizes, each to the value of £1 or more, in post-graduate, minor limit, mid-limit, limit and open classes may not compete at sanction shows. Later in this chapter I will list the various classes, because they must appear mystifying to the beginner.

It will be seen that the local sanction show affords a wonderful opportunity for the novice dog and novice owner. Another point to be remembered is that there is no stated time before which dogs may not be taken home from the show. This is a great help to the owner with a young dog, who would quickly tire should the judging run a bit late. Dogs do not have to be benched at sanction shows. This means that they do not have to be put on a bench such as is provided for them at the bigger shows, and for which the exhibitor has to pay usually 3*s*. per dog per show, however many classes he may be entered in. I find that inexperienced terriers hate being fastened up short to a bench with a collar and chain, especially if they have been trained, as I have previously suggested, to rest in a travelling box. Whenever I take a new dog or puppy—a novice dog—to an unbenched show I always take him in his travelling box, in which he may rest quietly until judging starts and in between classes. When he is required he comes out of his 'home' alert, tail wagging and showing for all he is worth.

Limited shows

This means limited to the members of the club, so that if you have already paid the subscription to become a member of a canine society, and they run a sanction and a limited show during any one year, your dogs are eligible for either or both shows. At limited shows dogs that have won a Challenge Certificate may not compete, but the competition is much stronger than at the sanction shows, and classes are usually put on for many separate breeds. Thus, if Wire Fox Terriers are scheduled separately, naturally it is less difficult to win in such classes than when one has to enter, say, in Any Variety terrier classes.

Open shows

Open to all comers. Here, the competition is even keener than at limited shows. Often about six to eight classes are provided for both Wire and Smooth Fox Terriers, as well as several mixed classes. The competition is usually particularly keen at open shows because dogs may be entered regardless of how many prizes they may have won and regardless of whether they are champions. Many champions of various breeds are shown at open shows, thus making the competition very keen and difficult to win the coveted title of Best-of-Breed or Best-in-Show.

If the novice exhibitor really desires to see how his stock compares with the well-known winning dogs the open show is the ideal show to attend. There the trimming may be compared with that of the experts, and a great deal can also be learnt about handling.

Puppy and novice classes are quite often scheduled at open shows. These are especially for the beginner or novice exhibitor, so until you are experienced in the art of entering and showing your dog my advice is to enter in the two lowest classes to begin with.

Open shows usually have benches for each exhibit, and all dogs must remain on their benches, each to his own number, all day, excepting when being exercised or judged. This is a Kennel Club rule. The time that you may take your dog home at the end of the show is printed in the schedule and all dogs must stay until the given time, which is another Kennel Club rule. These rules, and many others, are printed at the back of all show schedules, which, as I have said before, will be sent to you by the show secretary if required.

As well as the many breed classes at open shows, there is a number of Any Variety classes, usually judged by a different judge. The Variety judge is invariably a well-known 'all-rounder'—that is, an experienced judge of most breeds—whereas the breed classes are usually judged by a specialist in his or her particular breed.

Championship shows

These are the largest and most important shows of all. There are usually about ten or twelve classes for Wire Fox Terriers at championship shows, and the same number for the Smooth variety. At most of the championship shows, Challenge Certificates are offered for each breed, one for the best dog, and one for the best bitch. These are the most coveted wins of all. The Award Cards are large green and white cards on which is written:

'I am clearly of opinion that "Pride and Joy" is of such outstanding merit as to be worthy of the title of champion.'

Therefore it can be easily understood that the judge takes the signing of this card very seriously indeed.

The Award Card, which is awarded in the ring, is not the official certificate. This follows later by post from the Kennel Club, in about thirty days' time, when the win has been ratified and everything to do with the certificate winner has been found to be in order.

When the dog has won his third such 'green ticket' the Kennel Club, in due course, will send the champion's certificate to the owner. This comes automatically and should not be applied for.

To win a Challenge Certificate the exhibit has to be adjudged the best of its sex in its breed at the championship show.

For a dog to become a champion he must win three Challenge Certificates under three different championship show judges.

All the best dogs in the country are shown continuously at championship shows, and the competition is always very strong indeed.

Many of the top winning Wire and Smooth Fox Terriers are prepared and handled at the championship shows by professional handlers. This sporting body of men keep the standard of presentation and handling extremely high. Therefore it must be realized that to win a Challenge Certificate is a great achievement. Of course, to be able to write 'Champion' in front of a dog's name greatly enhances the value of the dog, and also that of its progeny.

To the beginner, the classification of a show may appear very puzzling. The following list of the definition of the various classes at championship shows is issued by the Kennel Club, and printed in all the show schedules. From these there must be one or two classes very suitable for anyone's dog.

Puppies under six calendar months of age are not eligible for exhibition, except at exemption shows.

PUPPY—For dogs of six and not exceeding twelve calendar months of age on the first day of the show.

JUNIOR—For dogs of six and not exceeding eighteen calendar months of age on the first day of the show.

MAIDEN—For dogs which have not won a Challenge Certificate or a 1st prize of the value of £1 or more (puppy and special puppy classes excepted).

NOVICE—For dogs which have not won a Challenge Certificate or three or more 1st prizes each of the value of £1 or more (puppy and special puppy classes excepted).

DÉBUTANT—For dogs which have not won a Challenge Certificate or a 1st prize of the value of £2 or more (puppy and special puppy classes excepted).

UNDERGRADUATE—For dogs which have not won a Challenge Certificate or three or more 1st prizes each of the value of £2 or more (puppy and special puppy classes excepted).

GRADUATE—For dogs which have not won a Challenge Certificate or four or more 1st prizes each of the value of £2 or more in graduate, post-graduate, minor limit, mid limit, limit, and open classes, whether restricted or not.

POST GRADUATE—For dogs which have not won a Challenge Certificate or five or more 1st prizes each of the value of £2 or more in post-graduate, minor limit, mid-limit, limit and open classes, whether restricted or not.

MINOR LIMIT—For dogs which have not won two Challenge Certificates or three or more 1st prizes in all, each of the value of £2 or more, in minor limit, mid-limit, limit and open classes, confined to the breed, whether restricted or not, at shows where Challenge Certificates were offered for the breed.

MID-LIMIT—For dogs which have not won three Challenge Certificates or five or more 1st prizes in all, each of the value of £2 or more in mid-limit, limit and open classes, confined to the breed, whether restricted or not, at shows where Challenge Certificates were offered for the breed.

LIMIT—For dogs which have not won three Challenge Certificates under three different judges or seven or more 1st prizes in all, each of the value of £2 or more, in limit and open classes, confined to the breed, whether restricted or not, at shows where Challenge Certificates were offered for the breed.

OPEN—For all dogs. If confined to a breed or variety, for all dogs of that breed or variety.

BRACE—For two exhibits (either sex or mixed) of one breed or variety belonging to the same exhibitor, each exhibit having been entered in some class other than brace or team.

Prize money at the championship shows is usually £2 for 1st prize, £1 for 2nd prize and 10*s*. for 3rd prize. At open shows the prize money is £1 1st, 10*s*. 2nd and 5*s*. 3rd.

Wins in Variety classes, i.e. classes in which different breeds are shown together, do not count when entering the dog in its separate breed classes. Also, do remember the term 'novice' applies to the dog and not to the owner. An owner of many champions can still enter a new dog (a novice) in any of the junior classes.

Very few, even the largest, of our championship shows schedule all these classes at the one show. The organizers and the committee of the show decide which of these classes in their opinion are most likely to be filled by exhibits, and they prepare and issue their schedule accordingly.

There are numerous all-breed championship shows held each year awarding Challenge Certificates for the various breeds, including Wire and Smooth-coated Fox Terriers, as follows:

February	Cruft's (two days), London
March	Manchester
April	Cheltenham
May	Bath
	Glasgow
	National Terrier Club Show at Leicester
	Leeds
June	Chester
	Three Counties Show at Malvern
	Blackpool (three days)
July	Windsor
	Paignton
	Cardiff
August	Leicester
	Dumfries
September	Hove
	City of Birmingham

	Bournemouth
	Belfast
October	The Scottish Kennel Club's Show at Edinburgh
November	The Ladies' Kennel Association Show in London
December	Richmond Dog Show Society, London

The great Cruft's Show, the largest dog show in the world, is held annually at Olympia. To win a prize at Cruft's is considered of great importance not only in the dog world but also among the general public. Cruft's is definitely a household word. Fanciers from many parts of the world visit Cruft's each year and the ringside is always packed with knowledgeable breeders.

Besides these general championship shows there are the specialized Fox Terrier Club championship shows which are only for the two varieties of Fox Terrier. They are held in various parts of the country. The Fox Terrier Club's Show is considered very important by terrier fanciers and the much coveted Gold Challenge Cups are awarded each year for the Best-in-Show for both Wires and Smooths. The Wire Fox Terrier Association holds its championship show annually, where many valuable cups are awarded. Then there is the Smooth Fox Terrier Club Show which specializes in the Smooth variety only. The Northumberland and Durham Fox Terrier Club and the Birmingham and District Fox Terrier Club cater for both coats.

A very important one-day championship show is the National Terrier Club's, which holds an excellent show each year, usually at Leicester, especially for all breeds of terrier.

Having decided upon the show (by looking in your weekly dog paper) at which you wish to show your dog, write to the secretary of the canine society running the show in question for a schedule. This he will gladly send to you. Select which classes you think most suitable for the dog or bitch you wish to show, and fill in the entry form very carefully. When this is completed post it together with the entry fees to the secretary of the show. If the show is to be benched there is usually a benching fee of 3*s*. for each dog, so send this too. All instructions are contained in the schedule, which accordingly should be read very carefully. Entries usually close about a month before a championship show and about a fortnight previous to an open show. For the smaller shows the time is less.

About a week before the day of the show the secretary will send you a pass with the number of your exhibit written on it. This must be taken to the show with you.

The great day of your very first show arrives. This is most exciting. I have been, literally, to hundreds of shows over the years, and even now, as soon as I awaken on a show day, I immediately become excited and full of nervous energy. It is very important to have everything ready the night before the show. There are quite a number of things that have to be taken to the show for the final preparation of your terrier, and these include the following, which I advise should be packed in a separate bag or small suitcase:

A tin of soft chalk, a lump of hard chalk, white Vaseline, Brylcreem, a comb, two brushes, one for the whitening and the other for the final polishing. A dish or small bowl for food and water.

Food must also be taken if it is an all-day show, but the water, which you will need for the dog to drink, or for damping or washing the legs, should they become soiled, can be obtained at the show. A show lead must also be taken. The one I use is very thin, not more than a quarter of an inch in thickness, with a patent sliding noose. If the show is benched a piece of blanket is needed for the comfort of your dog while sitting on the bench. Also, a collar and bench chain, not a leather lead, is a 'must'. If you do not possess one they can be purchased at the show. There are usually stalls selling such things at the open shows, and always at championship shows.

Another small item to take is a clip, or a safety-pin, with which to pin your number on to your frock, overall or coat. This can also be bought at the show.

Do not feed the dog before travelling to a show; it often causes sickness.

Be careful to start in plenty of time and make your way slowly to the venue. This prevents one flustering and bustling one's way to the entrance, which probably not only upsets the owner but also the dog. Dogs quickly sense when their masters are harried and upset in any way.

At the entrance to the show the pass you will have received from the show secretary will be inspected, and you will be allowed inside.

The next thing to do is to find your bench by the number given on your pass. Fix your terrier safely on his bench with the collar and bench chain, and then unpack your bits and pieces, such as your outer coat and your food, and place these under the bench for safe keeping.

Usually all shows provide a special place for cleaning and 'chalking up' the dogs. It may be a small room in an indoor show, or, if an outdoor show, a small tent. Several tables are commonly supplied there especially for the exhibitors. Find this place, and 'stake a claim' to a small part of one of these tables. But, better still, take a folding table with you.

If the show is one of the smaller ones, without benching, as I have said before, it makes things very much easier if you take your terrier to the show in a travelling box. Once he has passed the vet he can be returned to his box, and the box should then be placed in the room or tent provided for chalking up. I use paper in the boxes and a piece of blanket. This does not make such a mess as woodwool and sawdust do when the dog comes out of the box.

After you have exercised your terrier, stand him on the table and prepare him for showing, as described in a previous chapter. Allow yourself plenty of time for this final titivating and keep your dog standing on the table until it is time for him to be taken into the ring.

Next, buy a catalogue, which gives the names of all the entries, dates of birth, the sires and dams and their breeders and owners. These are well worth studying after the show, to see how the winners are bred.

One of the most important things to do before it is time to enter the ring for judging is to be certain that your terrier has been exercised sufficiently. It is sheer bad management or thoughtlessness, even laziness, that forces a dog to relieve himself during the short time he is actually being shown.

Judging seldom starts before eleven o'clock, and the limited and sanction shows sometimes do not start before 2.30 p.m. Surely this allows sufficient time for this all-important exercise. Remember, dogs that are uncomfortable move badly, often straddling their hind legs and roaching their backs.

When the time has arrived for the judging of the class in which you have entered, your number will be called out, or the steward will come and tell you that your class is about to be judged. He will also tell you in which ring the judging of terriers is to take place. Do therefore be on the alert. Believe me, it is most annoying to miss one's class. This is the moment you have been waiting for. Put the final touches to your dog, letting him shake himself before going into the ring. It gives a bad impression if he does this in the ring, making a cloud of chalk. Also, please brush off any chalk that you may have on your own clothing and shoes before going into the arena. Once you are there, your pre-entry 'butterflies in your tummy' will vanish, and you will find that all your time is taken up with getting the best out of your terrier.

If you have not already been given your ring number the ring steward will present you with it directly you arrive in the ring. This has to be fixed on to your coat or whatever you may be wearing, so that the onlookers can easily see it. This enables them to find your number in the catalogue and so learn the breeding of your terrier. While holding the lead of your exhibit in one hand, and trying to fix the card with the other, you no doubt will wish that you had at least two pairs of hands, so do not forget to have your special safety-pin clip already pinned into position before you take your dog into the ring. I realize that this advice is superfluous to the experienced exhibitor, but over and over again I see people looking helpless when trying to do these two things at once, simply because they have forgotten their pin.

The ring steward will tell you where to stand with your dog, which no doubt will be with the other newcomers, in the class on one side of the ring.

The judge will then come and look at the dogs while they are standing in a row. This is when all the time and patience you have spent on your terrier really pay dividends. Make sure that he is standing well, with his front feet together, his hind legs well out behind him and his head and tail high. Do not let him snap or snarl at the dog standing next to him. This completely spoils his expression, and makes him a perfect nuisance to all the other exhibitors.

In all probability, the judge will next ask you to place your dog on the table especially provided for the purpose inside the ring. He will then examine the dog most thoroughly, looking at his head, ears, eyes, teeth, length of neck, shoulder placement, bone, feet, spring of ribs, length of back, hind quarters and set-on of the tail. Also the texture of coat and condition. He will be able to judge his size better when all the dogs are standing together.

When he has completed this examination he will ask you to 'move' your dog. This simply means walking him away from the judge and back to him in a straight line. Be careful to have the lead well up under the dog's throat, neither too loose nor too tight. If it is too loose the dog may easily pull sideways on the lead, and if it is too tight he will 'mince along' with his front legs, which gives the impression that his legs do not touch the ground. Practice will have taught you the correct pace at which you yourself must walk so that your terrier moves smoothly and easily, and does not pull either forward or backward on the lead.

After the judge has seen the dog moving sufficiently to decide whether he is sound or not, he will ask you to stand on one side of the ring again, where you were to start with. Keep your eye on the judge, being very careful to have your terrier looking his best, right on his toes and very alert, while the judge compares all the terriers up and down the line. He is now able to study their outlines, and compare the size and style of each exhibit. Eventually he will probably ask about four of the terriers he likes best to come forward into the centre of the ring. Then, with perhaps another look or two, to be quite sure that he has not missed a carefully hidden fault, he will place them 1st, 2nd, 3rd and reserve, mark his judging book, and the class is over. The high opinion you had of your terrier has either been confirmed or your hopes have been dashed. Whichever way it goes for you, you must have a 'poker face', and learn to take the judge's decision graciously. There is always another day, and 'the last thing to be desired is success so complete as to put an end to trying'. This last quotation I stuck in my scrapbook in 1931 and have read it many times since. Another one I like very much is, 'Retain the faith that the best is yet to be'.

So good luck to your showing, and may you obtain as much pleasure and satisfaction from your efforts in the show, and from meeting so many friends with like interests, as I have enjoyed over the years.

CHAPTER II

THE KENNEL CLUB

THE reader who has persevered through this book will be aware of the frequent references to the control and guidance of the Kennel Club for the well-being and protection of the pedigree dog.

To anybody new to the exhibition and breeding of pedigree stock such an interested authority is of immense value. Not only does the Kennel Club lay down strict rules of conduct, without which the complete show system would become chaotic, but it also gives generous and prompt help to any breeder who is in need of assistance. On many occasions I have obtained valuable information and advice when I have been confronted with one of the many problems in connection with pedigree research, export pedigrees, change of names and so on.

The records maintained by the club of all breeds, and details of awards in the show ring, over a great many years are very extensive and complete. How complete they are is evident from the meticulous care with which all show successes are scrutinized. Any exhibitor of experience will have had sharp reminders of this scrutiny, which repeatedly leads to disqualifications of those exhibits which do not conform with the regulations governing the show classifications.

Quite unwittingly an exhibitor may enter a dog in a class for which he is not eligible, perhaps by reason of former wins or age limits. The dog might even win the class, only for the owner to receive a polite note from the Kennel Club pointing out its ineligibility and consequent disqualification. One might feel annoyed and disappointed by this treatment, but a little thought will soon make one realize how valuable to the whole organization of the show business is the constant care and control exercised by the Club.

Familiarity breeds contempt, and dog fanciers as a whole tend to regard the Kennel Club somewhat as the general public does the police—a force to be respected and thankful for, even though one does sometimes find oneself on the wrong side of the law.

The Kennel Club is regarded as the premier canine organization in the world. Other countries have followed their example, but none, I believe, has achieved as high a standard of efficiency and integrity as our own Kennel Club of such international fame.

I am indebted to the Secretary of the Kennel Club for permission to quote the following short history of the Club:

THE KENNEL CLUB

The middle of the nineteenth century found the prosperous and leisured Victorians with a passion for exhibitions and 'instructive entertainments'. The development of the railways brought the whole of the country within reach of the Great Exhibition of 1851, housed in what was to become the Crystal Palace; and during the following years a series of exhibitions of widely varying character was held.

The first organized dog show was held in the Town Hall, Newcastle-on-Tyne on the 28th and 29th June, 1859. The Show was organized by Messrs Shorthose and Pape at the suggestion of Mr R. Brailsford and had sixty entries of Pointers and Setters. There were three judges to each class and one of the judges of Setters succeeded in taking first prize in Pointers, whilst one of the judges of the Pointer classes took first prize in Setters. Only one class was held for each breed at these early shows and the dogs were unidentified except by their kennel names; reference to the old catalogues reveals Mr Murrell's 'Spot' price £5,000 competing against Mr Brown's 'Venus' price 22/-. Unfortunately, the name of the winner is not given.

The Calendar gives two shows each year for the next ten years and by this time, 1870, it was obvious that a controlling body was necessary to legislate in canine matters. The Crystal Palace Show was first held in 1870 by the National Dog Club and after the second show Mr S. E. Shirley, M.P. of Ettington, called together that Committee and a discussion ensued which resulted in twelve gentlemen meeting at No. 2 Albert Mansions, Victoria Street, London, in April, 1873 and this meeting marked the founding of the Kennel Club. One of the earliest undertakings of the newly formed Kennel Club was the compilation of a Stud Book, the Editor was Mr Frank

C. S. Pearce, the son of the Rev. Thomas Pearce, the well-known 'Idstone' of *The Field.*

The first volume of the Stud Book contained the records of shows from 1859 and to quote *The History of the Kennel Club* it was obvious that some system of distinctive nomenclature would have to be introduced to overcome the confusion arising out of 'quantities of Spots, Bobs, Bangs, Jets, Nettles, Vics, most of them insufficiently described and none of them being well-known dogs of the same name'.

In 1880 the Committee introduced a system of 'universal registration', which was strongly opposed, at first, but the advantage of reserving the use of a name for a dog was quickly seen and accepted. Registration in 1880 was nothing more than the registration of a name to avoid duplication in the Stud Book, the pedigree was of little importance and only came as an aid to identification at a later date.

The Committee formulated a code of ten rules, relating to dog shows, all of a simple character. It was announced that societies which adopted this code of rules for their shows would be 'recognized' and the winners at their shows would be eligible for the Stud Book. In 1875, the Committee decided to disqualify dogs which were exhibited at unrecognized shows but this rule was not enforced for some years.

The Kennel Gazette was first published in 1880 and has continued as a monthly publication from that date. The Stud Book and Calendar has been published annually, this year's volume is the ninety-first.

The Committee introduced many new rules and regulations and after a few years evolved a system of government which proved so sound that it was adopted as the basis of many overseas Kennel Clubs. At the end of the century the character of dog showing had so improved that about half of the exhibitors were women, and members of the Royal Family were showing dogs regularly. His Royal Highness the Prince of Wales was a staunch supporter of the movement to prevent the cropping of dogs' ears and from April 9th, 1898, such dogs have been ineligible for competition.

In 1900 nearly thirty championship shows were held and the smaller informal shows were becoming more and more popular. The policy of the Committee was to keep rules and restrictions to a minimum and shows were 'recognized', 'licenced' or 'sanctioned' providing the executive of the show agreed to adopt the Kennel Club show regulations. The guarantors of a show signed an undertaking (and still do) 'to hold and conduct the show under and in accordance with the Rules and Regulations of the Kennel Club etc.'

Each decade, the number of dogs registered at the Kennel Club has doubled. In 1963, 141,317 dogs were registered and 5,132 were exported, about one in every 27, a remarkable figure. 2,059 shows of all types were held in 1963, and 2,073 in 1964.

The Kennel Club acquired Cruft's Show during the 1939–45 war, after the death of Charles Cruft, and has held the Show in February each year at Olympia since 1950. The entry each successive year has broken all records—in 1964 there were 8,277 dogs of different breeds at the Show.

From the above the beginner will become aware of the prestige of the Kennel Club and the importance of its regulations, which must be adhered to in minute detail. Membership of the club is by nominations and elections. The main constitution of the club is exclusively male, but there is an important and influential ladies' branch. The wives of great men have, through history, brought pressure to bear at crucial moments, and I have no doubt that the ladies' branch, although only considered a social body, has a great influence upon the politics and legislation of the governing males.

Members of the Club, including the ladies' branch, enjoy free issues of the monthly publication *The Kennel Gazette*, and also receive the voluminous *Kennel Club Stud Book*, which is published annually. However, both these publications can be purchased by non-members. They contain a great deal of information respecting registrations, show awards, pedigrees, junior warrant winners, forthcoming shows, export sales, transfers and so on. They are a 'must' for the serious breeder and exhibitor.

The Kennel Club occupies a handsome modern building at 1–4 Clarges Street, Piccadilly, London W.1., where members may enjoy attractively equipped lounges, conference rooms, dining-rooms, library, cocktail bar, etc. From this it is clear that membership of the club is much to be desired, both from a social and informative and helpful aspect.

The regulations of the Kennel Club comprise a formidable list of documents, all of which cannot be included here. Reference has previously been made to some of the regulations affecting the shows. The following are of particular interest to the breeder-exhibitor, and relate to exhibition at shows:

All dogs exhibited at the show must be registered at the Kennel Club in accordance with the provisions of Kennel Club Rule 5. Forms for this purpose will be supplied on application to the Secretary, the Kennel Club, or to the secretary or manager of the show.

Disqualification and forfeit of prizes. A dog is liable to be disqualified by the Committee of the Kennel Club from winning any award, whether an objection has been lodged or not, if proved among other things to have been:

(*a*) Exhibited at an unrecognized show.

(*b*) Exhibited by a person disqualified or suspended under Rules 15, 17 or 18 as from the date of the charge having been lodged at the Kennel Club and for the period of the disqualification or suspension.

(*c*) Entered after the date fixed for closing entries or in a manner not complying with the classification in the schedule, or not in accordance with Kennel Club rules and regulations.

(*d*) (i) Totally blind; (ii) afflicted with a tendency to reproduce any condition which, in the opinion of the Committee of the Kennel Club, is hereditary and deleterious; (iii) defective in hearing; (iv) castrated, spayed, or otherwise prevented from breeding as the result of a surgical operation; (v) not entire. (An entire dog is one which has two apparently normal testicles both descended and in the scrotum.)

(*e*) Suffering from an infectious or contagious disease, or prepared for exhibition in contravention of regulations.

(*f*) Exhibited for competition or handled in the ring by a judge of dogs of the show, except where a judge is appointed in an emergency.

(*g*) Owned by the judge within a period of twelve months or, if a puppy, bred by him, except that this shall not apply to a substitute judge, if the judge of the breed be changed after the closing of entries.

(*h*) Handled at a show, boarded or given attention within twelve months by the scheduled judge of the breed of the show.

(*i*) Absent from its bench contrary to the provisions of Regulations 19(*a*) and (*b*).
(*j*) Not duly registered or transferred at the Kennel Club.
(*k*) The progeny resulting from the artificial fertilization of a bitch, unless prior permission was obtained from the Committee of the Kennel Club, and such conditions were complied with as it is considered advisable to prevent the possibility of fraud.

If a dog is entered in a class for which it is ineligible and is not withdrawn from competition before commencement of the judging of the class in question the dog shall be disqualified. All such cases must be reported by the exhibitor to the show secretary, and by the latter to the Kennel Club, and the exhibitor may be fined at the discretion of the Committee of the Kennel Club.

The owner of a dog disqualified for any of the above reasons is liable to forfeit all entry fees paid for, and all prize money won by, such dog.

The Committee of the Kennel Club shall have the power to inflict fines upon exhibitors who shall have made errors in their entry forms or breaches of show regulations.

In the event of such fines not being paid within the time stipulated by the Committee, the exhibitor or exhibitors may, at the discretion of the Committee, be dealt with as if a complaint under Kennel Club Rule 17 had been lodged against him or them and proved to the satisfaction of the Committee.

All entries must be made on the entry form enclosed with this schedule; they must be written entirely in ink or indelible pencil; only one exhibitor shall enter on one form.

A copy of the K.C. Rules and Regulations, as supplied by the Kennel Club, must be accessible at the show to every exhibitor.

No dog under six months of age can be exhibited.

No person may interfere in any way with the benching as erected in accordance with Regulation 18 of the Kennel Club Show Regulations and in particular the partitions as erected between one pen and another must not be removed under any circumstances. Printed cards 'For Sale' and 'Not for Competition' may be placed over benches, and in the case of a dog entered 'Not for Competition' the name of the dog may also be displayed, but in no case may the lettering be larger than 1 inch. Exhibitors may display a board or card over their benches containing the name of their Kennel, the name and address of the owner of the Kennel, the telephone number, but no other information whatsoever. The size of the board must not exceed 8 in. × 5 in. and the lettering must be black on a white background.

No dog shall be absent from its bench except for the purpose of being prepared for exhibition, judged, exercised, or by order of the veterinary surgeon or show committee. After being prepared for exhibition the dog must be replaced on its bench. No dog shall be absent from its bench for the purpose of being exercised at any time for longer than fifteen minutes. No dog shall be allowed in a judging ring other than actual exhibits in the class being judged. All dogs when off their benches must be on a lead throughout the duration of the show except in the exercising area, when provided, or when competing in the ring in an Obedience Test. Prize cards, award cards, or rosettes may be displayed on the bench, but not in the ring.

Prefixes and affixes. The serious-minded breeder exhibitor no doubt, sooner or later, will be desirous of having his own prefix or affix. These are Kennel names applied before or after the individual dog's names. These names or trade marks must be approved by the Kennel Club. Once granted, the prefix or affix can be used only by the person who registers it. To do this, one has to apply to the Kennel Club for a special form to fill in, giving the name desired. The Kennel Club asks for three names to be submitted for their consideration, in order of preference. All names applied for are published in *The Kennel Gazette* each month. Above this list is printed 'The following list of prefixes and affixes applied for will be placed before the Committee and the words granted will be published in the reports of the meetings in *The Kennel Gazette*. Any person having an objection to any of the words listed should communicate with the Secretary within fourteen days'.

The cost of registering a prefix or affix is £3 and the maintenance fee is £1 per annum. The maintenance can be compounded for life by paying a fee of seven guineas, or, after ten years, five guineas.

We are advised to register all the puppies we breed. This is because the numbers of Challenge Certificates allocated to each breed are governed by the total of the breed's registrations each year. I, personally, do not register my puppies until they are old enough for me to decide which names will suit each puppy. Imagine calling a dog

'Superb', or a similar name, if it does not prove to be up to show standard, and no more superb than I am.

I am proud of my prefix, and do not let any terrier carry it unless I consider it good enough to enhance my Kennel name. Quite a number of breeders, when buying a puppy from me to introduce new blood into their own Kennel, repeatedly ask me if they may be allowed to register the puppy with their own prefix. On the other hand, quite a number of owners to my knowledge will not sell anything at all until it has been registered with their 'trade mark'. It is just a matter of opinion, I suppose.

Change of Name. Should you buy a dog that has already been registered you may add your own prefix/affix to its name, provided the dog in question has not been entered in the Kennel Club Stud Book.

For a dog to be entered in the Stud Book automatically he must win a 1st, 2nd or 3rd prize in limit or open classes at a championship show. The name of a dog which has qualified accordingly for entry in the Stud Book cannot be changed after thirty days have elapsed after the date of qualification.

I have found it most important to keep a list and the dates of dogs winning in either limit or open classes at championship shows. As an instance of how frustrating failure to do so can be, I mention an occasion when my young dog won a third prize in the limit class. Shortly after this he was sold for a large sum of money *provided* the new owner's world-famous prefix could be added to his registered name. I thought this in order when I accepted the offer, quite forgetting the miserable 3rd prize the dog had won. By this time he was winning Challenge Certificates. Then suddenly it dawned on me that about thirty-three days previously he had become eligible for a Stud Book number and another prefix could no longer be added. I cabled my client, who, in turn, telephoned the Kennel Club several times asking for his prefix to be added, but all to no avail. As this was a condition of the sale, naturally it fell through. So do keep a list of each dog's winnings, together with the date of the show.

Other dogs with certain qualifications may be entered in the Stud Book by their owners on the payment of a small fee. The qualifications for this type of entry being: (*a*) The sire or dam of a winner entered, or entitled to be entered, in the Stud Book; (*b*) A son or daughter of winners entered, or entitled to be entered, in the Stud Book.

Here I would like to give another piece of advice. Never be afraid of selling a good dog. Selling a bad dog does immeasurable harm not only to your name but also to the dog fraternity in general. I would rather sell a good dog at a fair price than a bad dog for a good price. This may sound smug and self-satisfied, but, believe me, it pays over and over again, because a satisfied customer always returns. The dog world is small, and bad publicity travels fast.

Junior Warrant. This is an award which can be won by a consistently winning young dog. To win a Junior Warrant, the dog must win twenty-five points before he is eighteen months of age. Three points are awarded for each 1st prize won at championship shows, and one point for each 1st prize won at open shows. Only wins in separate breed classes count. Wins in any variety classes therefore do not count at all. The Junior Warrant award has to be obtained from the Kennel Club by application on a special form, giving the list of winnings and the dates of the shows.

Breeding terms agreements. Some owners loan their bitches out on breeding terms as described in a previous chapter. The Kennel Club issues, for a fee of £1, a special form of agreement to be signed by both parties in such cases. It provides a contract for the 'Loan or use of Bitch'. If this contract is used the conditions of the breeding terms are officially recorded at the Kennel Club. If, later, there should be any dispute as to the fulfilment of the contract, then the Kennel Club becomes the final judge of the matter.

However, a great many friendly arrangements are made about breeding terms, without their being recorded at the Kennel Club. This practice, as I have said before in a previous chapter, is, in my opinion, very unwise, as it so often leads to unpleasantness.

Export requirements. Before any dog can be exported it is necessary to obtain a certificate from the Kennel Club, giving its pedigree. This is for the satisfaction and protection of the purchaser who, with such a pedigree in his possession, can register the dog with his own kennel club. The Export Pedigree, as it is called, provides the warranty of the English Kennel Club which is recognized throughout the world. The present charge for an Export Pedigree is £2, or two pounds ten shillings, according to destination of the dog. The owner of a dog can have a Registration Card stamped by the Kennel Club 'not for export', and if this is done the Kennel Club will not issue an Export Pedigree.

In the case of a male dog, an additional Kennel Club form respecting monorchidism and cryptorchidism is necessary, certifying the animal's freedom from these defects. Before an Export Pedigree can be issued it is necessary for the dog to be examined by a qualified veterinary surgeon. Finally, a health certificate completes the requirements previous to export. This must be signed by the veterinary surgeon following examination of the dog within seven days of its leaving the country.

Breeder's Diploma. This is a certificate issued by the Kennel Club to the breeder of a dog or bitch which becomes a champion, whether or not the breeder is the owner of the animal when it qualifies for championship status. Many people cannot spare the time or the money required to attend championship shows regularly. Quite often such breeders sell their promising young stock to enable them to have the chances in the show ring which they deserve.

These diplomas are, and should be, proud possessions. They are not awarded automatically by the Kennel Club, and must be written for, giving the three championship shows and the dates at which the three qualifying Challenge Certificates were won. There is no special form for this.

Always remember that the Kennel Club is there to help and not to hinder. I, personally, cannot speak too highly of their administration. Any enquiry is courteously answered. To call there for any reason at all, to me, is a pleasure, and makes me realize how vast and important the breeding and exhibition of pedigree dogs has become.

CHAPTER 12

EVERYDAY AILMENTS AND HOME NURSING

IT HAS been suggested that I write a chapter on common ailments and home nursing. Having given this much thought, I have decided that as drugs of today are constantly changing, and being improved upon, it would be most unwise for me to attempt to offer advice as to the use of them.

Here I particularly wish to stress the great importance of calling in a veterinary surgeon as quickly as possible, because it is most difficult for anyone outside the veterinary profession to diagnose what is wrong with an animal. No two dogs show the same symptoms, even in the same illness, and, therefore, immediately I find one of my dogs to be off colour, even after so many years of experience, I never fail to ask for my veterinary surgeon's help. He, being a small-animal specialist, no doubt sees and examines more sick dogs in a week than I do in a decade. Personally, I feel that, with the advancement of veterinary research and with the veterinary surgeon's long and comprehensive training and experience at our disposal, we should go on our bended knees and say a big 'Thank you' for the help so readily available.

However conscientiously and careful one is in looking after and caring for animals, there comes a time when little things go wrong and accidents happen. As I have mentioned previously, it is surprising the number of things Fox Terriers can find with which to amuse themselves, sometimes with harmful results.

Many years ago a Wire puppy I had sold was staying in a boarding kennel for a holiday. He was being exercised in a run enclosed with medium-sized-mesh wire netting, next to a similarly enclosed run, but with no boards in between, as I have advised previously. The poor unfortunate puppy, then about six months of age, pushed his front leg through the netting, trying to play with the much larger dogs in the next run. His leg was literally torn to pieces, the flesh being ripped to the bone. Although veterinary assistance was called, and expert treatment given, the Fox Terrier puppy died, no doubt from shock as the main cause.

However small the accident, shock should be treated for straight away by applying heat, either with an infra-red lamp placed over the patient, or with the aid of hot-water bottles. Peace and quiet, too, are essential and a mild sedative may be necessary to bring relaxation. I personally find that when a dog needs sleep it is advisable to keep him in complete darkness, either by covering the kennel with a dark blanket or by drawing dark curtains across the windows. Dogs like being in darkness when feeling ill, they seem to feel so much more secure. I suppose this is a relic of their wild state when they no doubt made their homes in an earth, similar to those used by foxes.

A very important point to remember is that should your dog be unfortunate enough to meet with an accident, do not pull him about trying to find the injury. Keep the dog quiet and warm and call in the veterinary surgeon. My veterinary surgeon insists that many lives are lost by too much handling in the first twenty-four hours after an injury. So leave the dog alone, and keep the patient as quiet as possible until the veterinary surgeon arrives. Do not handle the patient more than is absolutely necessary. Pushing and pulling about can cause untold harm.

Home nursing of dogs, as with children, requires a good deal of common sense. One of the first essentials is to make sure that the patient is kept clean, warm, comfortable and quiet.

Should there be any discharge from the eyes or nose, this should be carefully wiped away with cotton wool, and the eyes bathed with a soothing lotion. I personally do not care to use ointment for the eyes, because any dust or dirt is inclined to stick to it, and can easily set up inflammation.

The inside of the mouth and gums must be kept clean and fresh by wiping the gums and teeth with cotton wool soaked in a weak solution of a mild antiseptic.

If any excreta becomes stuck to the rectum this should be gently removed with cotton wool and the soiled parts smeared with olive oil or Vaseline to prevent soreness. Dirty parts caused by diarrhoea must be washed with warm soapy water, carefully dried and dusted with powder.

During convalescence a gentle brushing and combing, once a day, helps to give the animal a sense of well-being. He enjoys the little extra attention bestowed upon him, but care should be taken that he does not become over-tired by this grooming. Long periods of peace and quiet are very necessary in illness.

Persuading the patient to take food during illness and convalescence is often a great problem. My experience has been that a little nourishment taken voluntarily does far more good than food forcibly administered. Food offered should be made to smell temptingly. Cooked meat, minced or cut up finely, should be placed under a hot grill for a few minutes. This brings out the smell and makes the food more inviting. Rabbit treated in this way will often prove palatable. Tinned salmon, which has quite a strong appetizing smell, sometimes is acceptable. A little and often is a good maxim to remember. Always bear in mind, too, that it is the food which a dog digests which does good, not bulk, which so often proves indigestible and which may cause diarrhoea or other stomach troubles.

In the old days diarrhoea was treated by feeding the whites of eggs, arrowroot, cornflour, Slippery Elm and suchlike foods. Nowadays there are many drugs which help considerably, about which the veterinary surgeon will advise.

The great thing to remember in home nursing is to treat each symptom as it appears, and to find its cause.

Ears need to be watched carefully for signs of ear canker, which is a term commonly in use for excessive wax in the ear. If the canker is severe there is a dirty reddish-brown discharge. This discharge gives off a nasty smell, and the dog carries his head on one side and scratches the outside of the ear continually. The only thing one can do to help this is to clean the inside of the ear carefully with cotton wool on an orange stick, being careful not to damage this very sensitive organ. Also use a lotion on the outside of the ear to soothe the scratched skin. Should cleansing the inside of the ear not prove effective, I strongly advise asking your veterinary surgeon for treatment.

Cysts sometimes appear between the toes, which the dog licks continually until they are broken. Personally, I have had little or no experience of these, but a number of people have telephoned me asking what to do with these irritating things. I have suggested, when these occur, bathing with hot water until the cysts have broken, after which an antiseptic lotion should be applied.

Parasites are easily dispensed with. Ordinary dog fleas can be killed by dusting with any of the proprietary powders. I have been told that it is sufficient for the powder to be placed in between the point of the shoulder blades, because all fleas pass over that part of the body once a day. But I am afraid this is an old wives' tale! My husband insists that he has had dozens and dozens of fleas on his own dog—a Foxhound—and that they could be seen running about on the hound's stomach, where they appeared to concentrate. In this case powder would be of no use as it would not stay on long enough to kill the fleas. He used my old-fashioned remedy of paraffin oil and milk with excellent results.

Lice can be easily recognized as they are a dirty white in colour. They do not jump or run like fleas, but stick their heads tightly into the skin. They appear in clusters, quite often in the crevices of the ears, and quickly set up skin irritation. Many years ago a litter of puppies I bred had them. I was using straw as bedding at the time, and I have always thought they came into the kennel with the straw. I find that a very effective way to be rid of these pests is to use a mixture of equal parts of warm cows'-milk and paraffin oil. Apply with a toothbrush and then wash the affected parts with warm water, and dry thoroughly. This does not harm even the most sensitive of skins and can safely be used for very young puppies. This remedy was given to me when I was a girl by an old rat-catcher friend of mine.

Ticks are big fat, whitish bugs. They have heads with which they suck the blood from the host. I use the same paraffin mixture for these, and the pull them out with a pair of tweezers, applying iodine to the resultant sore.

Worms have been mentioned in previous chapters, when I advised worming puppies at six weeks of age and also brood bitches when in whelp. In spite of worming bitches during pregnancy the puppies are invariably born with worms, and all dogs seem to need treatment for them from time to time. There are many well-known proprietary pills on the market for removal of the common round worms. These pills are easy to administer as the dog requires no fasting previously nor an aperient afterwards.

Tape worms are more difficult to get rid of, and often need two or three doses to effect a cure. These can easily be recognized by the dog passing flat tape-like segments of worms in its excreta. In difficult cases I obtain advice and the necessary pills from a veterinary surgeon.

A dog suffering from worms shows lack of condition in its coat. Worms quite often cause spasmodic feeding, resulting in the dog losing weight and looking out of sorts generally. Therefore I advise worming puppies twice or three times during their first six months, and perhaps twice a year as a general rule afterwards.

'Riff'. This is a term used among terrier people to describe a skin trouble which completely turns the skin, and also the hair, a pink colour. It usually starts on the inside of the hind legs and underneath the front legs. The condition seems to me to be contagious and quickly spreads from one dog to another. Absolute cleanliness of kennels is the first essential in clearing away this disfiguring complaint. Wash the pink parts with warm soapy water every day, say for one week, dry thoroughly, and apply a lotion consisting of camomile and surgical spirit. The hair does not fall off as in so many skin troubles, and the pink hair eventually grows out with constant cleanliness and the use of dry clean bedding and kennels.

Travel sickness is very common in dogs, but, like children, they usually grow out of it. It is, of course, unwise to feed a dog before a journey, or allow him to drink. Any of the human seasick pills, given the night before the journey, are certainly worth trying, but care should be taken not to give an overdose. The peculiar thing to me is that a dog is often sick going to a show, but seldom coming back. This points to the fact that once the stomach is empty, and the dog tired, he sleeps comfortably all the way home! Or is it that all the excitement is over and he is able to relax?

Ordinary cuts, scratches or abrasions of the skin should be treated either with iodine or with permanganate of potash crystals ground to a fine powder. Puncture wounds, probably caused through fighting, should be bathed with mild antiseptic water and not allowed to heal for at least a week. This is so that the wound heals from the inside and not just on the surface.

Do not forget 'that all pups eat muck'! There comes a time sooner or later when they eat too much muck! A very useful and harmless emetic is a small piece of ordinary washing soda, about the size of a pea, pushed down the throat. This is sufficient to make a terrier sick almost at once.

Here I wish to stress again the great importance of using the many very excellent inoculations now available. Anyone who has seen or nursed such distressing illnesses as distemper and hardpad will have all their dogs inoculated without fail.

Inoculations are almost infallible against distemper and hardpad, also canine hepatitis. There are excellent inoculations too for kidney trouble and jaundice, but these latter need a 'booster' injection every year.

Anybody starting a Kennel should consult a veterinary surgeon at the outset to be informed of the many preventive inoculations available. Prevention is better than cure, and I cannot speak too highly of the efficacy of these modern aids to health.

With these to hand, and with the application of common sense and care, nobody need be nervous about the well-being of their stock.

CHAPTER 13

DO'S AND DON'T'S

I AM AFRAID that, so far, some of what I have had to say in this book may have been technical and inevitably rather dull. Lest this should give the impression that there is little of fun and pleasure in the breeding of terriers and the show game, I will describe some of the incidents which have happened to me over the years and which I recall with interest and pleasure. These are but a few of the episodes I remember among a host of experiences in my lifetime hobby.

When in my teens, living at Wantage, my brother and I became acquainted with the late Tom Reveley, the well-known and outstanding dog and horse photographer of those days. We were constantly in demand to attract the dogs that he had to photograph, to ensure good pictures. In the Fox Terrier Club's Year Book of 1926 several of these pictures can be seen. By this means I became interested in Wire Fox Terriers, which have been my firm favourites ever since.

Among our friends at Wantage was Ted Robson, the race-horse trainer. On our way to the gallops one morning we met one of Mr Robson's grooms, a Mr Evans, who spent all his spare time at the Frilford Kennels, owned by Mr Bond. One day, Evans came to tell me that Mr Bond had said that, 'if Maggie has more than four puppies next week you may have one'. Maggie obligingly had five puppies and Evans brought the tiny day-old dog puppy which Mr Bond had given him for me to see, tucked inside his shirt next to his body. He reared the puppy by hand, taking it everywhere with him. I remember it so well, with even marked tan head, white blaze, attractively hound-marked and he seemed to me to be a little model. Eventually came the great day when Evans took the dog to his first championship show at the Crystal Palace. Apparently the youngster showed beautifully, and Evans told me that after moving the dog up and down, he dropped the lead and the dog stood like a statue. He won, and later became Ch. Frilford Felix (U.S.A.). A Mr Mattinson, a top professional handler of the day, sent a telegram to Evans asking what price he would take for Felix. Evans, I think trying to impress his small town friends, replied by telegram '£250', which was a lot of money in those days. Mattinson's answer came back promptly accepting the price, and so Felix was sold. Evans couldn't believe that he had actually contracted to sell his pet, and so hid the dog, where I don't know. Mattinson in the meantime had sold Felix to U.S.A. A lengthy court case followed, which Evans lost, paying costs. So Evans had to part with Felix much to his grief. Champion Felix was the first champion I knew.

Just before the war I purchased a young dog, whose sire was Ch. Triangle Jupiter. This ten-month-old puppy was sent to me on approval completely untrimmed. He was priced very reasonably, about £25, if I remember correctly. At this time, we were living at Bath, and Mr Leybourne-Popham of the then famous prefix 'Hunstrete' was a frequent visitor. He was most interested in any new dog or bitch I had. If they were good enough he coveted them and invariably persuaded me to part with them. This was easily done, because I always suffered from 'financial cramp', and he was a very rich man. This particular young dog I registered as Penda Planet, and I made up my mind to keep and to show him myself. I remember stripping him out, and how pleased I was with him when his make and shape were revealed.

A day or two later, Mr Popham telephoned saying that he would like to come to see the dogs the following day. I felt that if he were allowed to see Planet he most certainly would want to own him. Therefore, I put Planet into an enclosed travelling box and placed it behind a chair in the dining-room. Mr Popham looked at all the young stock in the Kennels and seemed on the point of departure when he decided that he had time to spare and would like a cup of coffee . . . in the dining-room. Planet was fast asleep when we entered, but on hearing us, he shook himself violently in his box, and this immediately aroused Mr Popham's curiosity. Planet was inspected very thoroughly while he was being moved up and down the lawn. Mr Popham obviously liked the dog very much and asked me if I would sell him. I said that I wanted to show him myself at the forthcoming Bristol Terrier Championship Show. We had a Bristol Terrier Show in those days, before the war.

Quite early the following morning the telephone rang. It was Mr Popham saying that he was unable to sleep for thinking about Planet, and could he please have him. He then offered far too much money for me to be able to refuse, and so Planet changed ownership, but Mr Popham allowed me to show the puppy at the Bristol show. He won well, and several well-known fanciers told me that I had sold a certain future champion. This grand young dog did not have the chance to prove or disprove this statement, because war was declared soon after this and championship shows ended during hostilities.

Later, Mr Popham asked me to sell Planet for him. Dogs were worth little or nothing at this time, and we were glad to accept twelve guineas for him to go to Ireland. Towards the end of the war, the dog was sent to America, I believe, but I never heard the end of him.

I managed to keep a few bitches during the war years, and also one dog called Penda Pompilius, sired by Ch. Talavera Romulus. It became increasingly difficult to keep terriers, as food became scarce and everything was rationed. Dog biscuits were at a premium, and I had to queue for hours each week for very small quantities of any food suitable to feed the dogs.

Penda Prize Packet was one of my best bitches then; she was lightly hound-marked with an even marked tan head, and I managed to breed several litters from her during the war.

When she was in whelp to Penda Pompilius on one occasion, a client called who particularly wanted a bitch puppy from her. I had no puppies left from her previous litter, and he persuaded me to sell him a bitch from her next litter, which was due in about a month. He insisted on my being paid in advance. This is when I learnt another lesson.

In due course the 'litter' arrived, comprising only one puppy—a bitch. She was a plain-looking puppy with an all-white body and an even-marked tan head. She had wonderful substance and a really hard coat, even at a very early age. I was loath to part with her but felt in honour bound to do so, but I made a vow never to sell an unborn puppy again.

She won quite well for her new owner, and he eventually sent her to one of the top professional handlers of the day. He won one reserve championship with her, but never a Challenge Certificate. Several times during her show career I tried to buy her back, but she was not for sale. When she was about four years old and had not been shown for quite nine months, I wrote again, as I particularly wanted to breed from her. I think by this time both the owner and the handler were completely disheartened and despaired of her ever becoming a champion. So they decided to let me buy her back. Not cheaply, by any means. She certainly did not look a show bitch when I brought her home! She was completely untrimmed, and had very long toenails and open feet.

I stripped her out completely, filed her toenails daily and exercised her in a cement run. After about ten weeks of this treatment the improvement was really amazing. I remember saying to my husband, 'Happy is too good a bitch to keep only as a brood bitch.' Shortly after this I entered her in a large open show. There were several well-filled Wire classes and the judge was Mr 'Solus' Bishop. I had never seen him before in my life, and he made me most indignant while examining my bitch—he nearly turned her upside down, feeling her coat all over. I asked him afterwards just what he was looking for. He said that her coat was so hard to the touch that he wanted to be certain that I had not used anything to harden it. As if I would.

She won all the Fox Terrier classes, then won the cup for the Best Terrier and later the cup for Best-in-Show.

Naturally, this greatly encouraged me, and I asked Mr T. Brampton, who is still one of our present-day top professional handlers, if he would try to make her a champion for me. He took her home with him and every time he showed her he always had her pure white coat in wonderful condition.

She won her first championship at Brighton under Her Grace the Duchess of Newcastle. This was my first Challenge Certificate, and was certainly one of my greatest thrills. I was staying with my friend Mrs Dorothy White at the time, and I remember waking her up in the middle of the night, after the show, asking her to switch on the light so that I could see my wonderful C.C. again!

'Happy' became a champion when she was five years old, and afterwards she bred me many winners, quite a number of which were exported to various parts of the world. She retained her beautiful lean head right to the end of her days.

Thurse

Ch. Crackwyn Cockspur (Born 1958)

Ch. Penda Worsboro Whistler (Born 1967)

Thurse

Thurse

Ch. Penda Peppermint (Born 1967)

Ch. Penda Tavatina (Born 1961)

Thurse

Well I remember the war years. The Kennel Club allowed only non-championship shows to be held, and, to prevent people travelling, shows were limited to entries from a twenty-five-mile radius. We were living at Bath during the war and I was thus able to show at Bristol, and one or two other shows near home.

The day after the evacuation from the Dunkirk beaches will always live in my memory. The Bristol Terrier Club held one of their shows on that day. We had previously enquired whether, having regard to the dreadful war situation, the show should be proceeded with. As we were told that dog shows were one of the means of keeping up public morale, we all turned up at the hall with our dogs.

Dog shows were few and far between in those days, and there was a tremendous entry, which was judged by the late Mr McDonald Daly. It was a lovely summer's day and Bristol was swarming with soldiers who had, a few hours previously, been taken off the Dunkirk beaches in the little ships and landed at Avonmouth. Bristol was an 'open house' to them, and a great many of them walked into the show, still in their mud-stained battle uniforms. They must have been amazed to find a dog show in progress at home while they and their comrades had been fighting their way out of their enemies' trap. However, they seemed interested in the show, and possibly the friendly and calm atmosphere helped them to relax.

Another time I remember during the war was that of the 'Baedeker' air raids on Bath. When the raids started my great fear was the danger of incendiary bombs falling on the wooden kennels. Consequently, as soon as the bombs started falling, my son and I rushed out to get the dogs into the house. On one occasion, just as we went outside we heard a plane flying very low overhead. Peter, my son, said, 'It's O.K., its too low to be one of theirs,' but, with that, as though the pilot had heard us, the bombs came whistling down. Before, or since, I don't think I ever moved so fast, and we both fell on top of each other in a heap in the kitchen. Anyway, we eventually got the dogs in and we spent the rest of the night with them, the cat and two evacuees in a cupboard under the stairs.

I think that one of the best characteristics of Fox Terriers is their sporting temperament, and I have had many which, while being highly bred show dogs, have been equally at home in the field hunting rats and rabbits, and even, on occasion, facing a fox or badger in their earths—and, believe me, it takes a very game dog to go down after a badger.

Terriers are also very intelligent, and I remember two bitches, called Marlene and Garbo, which I once had. These two worked out a first-class method of hunting rats which swarmed in some old warrens in a field behind our house. They soon learnt that indiscriminate chasing brought small reward, so they ended up by taking turns in digging out holes, while the other would sit and wait at a well-used bolt hole. As the rats shot out, they were easy meat for the waiting dog. A pretty good example of canine intelligence, I think.

Another endearing quality of Fox Terriers is their invariable gentleness with children. I am a great believer in giving children every opportunity to enjoy the companionship of animals, and I am sure their sense of responsibility and kindness is greatly increased as a result.

Of course, children must be taught how to care for their pets, such as keeping them clean, feeding and generally behaving properly towards them. It annoys me very much to see young children constantly squeezing puppies, or teasing and chasing them all day. No one can blame a harassed dog for eventually turning on a child after it has been goaded from morning till night for days on end.

Needless to say, my own family have been brought up with terriers around all the time, and I am afraid that on occasions they must have got a bit bored by hearing all about dogs at all times of the day and night!

Even so, my son's family are now carrying on the tradition and my daughter-in-law has been breeding terriers for some years now. I remember the first time she showed one of her dogs. This was a very useful bitch which had previously won at championship shows, and I think she fully expected to sweep the board. Anyway, she quickly learnt that it is not so easy to win, as she was sent out of the ring along with the rubbish.

However, Sheila persevered, and bred several good litters from the bitch before exporting her to America, where, I am informed, she still gives great pleasure to her new owners.

I particularly remember seeing the young men attending our radius shows during the war while they were on leave and, quite often, handling terriers in the ring. Les Greenhough, in his smart Air Force uniform, came to see us at Bath. Young Jimmy Butler, helping his father, Alf Butler, to prepare and show the beautiful 'war baby' Harley Fulgent, who impressed me very much. I remember sitting with the two terrier experts, 'Towyn' Edwards and Arthur Cartledge, watching this dog move—he went 'like a train'—and we all three agreed how much we would like a litter by him. This was not to be, however, for Harley Fulgent was sold to America, and, alas for all fanciers, he was sunk by enemy action while crossing the Atlantic.

Albert Langley, the well-known present-day handler, must have been a very good soldier, for I think I am right in saying that he became a sergeant-major in the Royal Warwickshire Regiment. When he returned after the war I believe he was with W. Brown-Cole for a while. One thing I am quite certain about is that he prepared and handled Mrs Dorothy White's very first champion Wyretex Wyns Princess. Since then, Albert has made well over fifty champions, far too many to mention individually, including two champion bitches of mine, both of which he also handled to Best-in-Show awards.

As I have advised previously, I always take my dogs to shows in their travelling boxes. How to tell you to fasten them, I really do not know. Years ago I always used padlock and keys to keep the dogs safe until I needed them for their particular class. On one occasion, when it was time for me to take my dog into the ring, I could not find the key anywhere. The stewards were frantically calling my number, and I and my friends were just as frantically trying to get my poor dog out of the box. It ended up by one of them running to tell the steward that I was just coming, while the others unscrewed the hasp and staple from the box, and so released my prisoner.

Next I decided that I would use just an ordinary fastener, no more locks and keys for me, I thought. The dog boxes were placed in a huge marquee at one particular open-air show I remember. It was an unbenched show. It always surprises me how far my dogs are from the Wire Fox Terrier ring. I went to the ringside to find out which class was being judged, and returned to my three boxes, only to find them wide open. The three dogs I could just see in the distance having a marvellous scamper, about three fields away. Children playing in the tent had obviously opened the boxes and the dogs were fully appreciating their unexpected freedom. Nowadays, I do not leave my dogs for any length of time unattended.

A story I like very much, which I am told is absolutely true, is about an enquiring novice who approached an exhibitor who was just completing the final preparation of his Wire Fox Terrier before entering the ring. The dog in question had an abundance of face and leg hair. The novice enquired: 'Do please tell me how you manage to grow such lovely leg hair. What should I put on my dog to help his to grow?' 'Leave it alone,' said the expert, 'just leave it alone.' Several months later, I am told, these two met again at another show, when the novice rushed up to the expert saying, 'Try as I will, I still cannot make leg hair grow, and I have used bottles and bottles of eau-de-Cologne'.

We have many laughs at shows both before and after the judging, but once in the ring the business becomes very serious indeed. There is a Kennel Club rule which says that speaking to the judge is not allowed. This is, of course, unless the judge asks a question.

Often, I have laughed at the story of one well-known exhibitor who is completely deaf. When asked for the age of the dog, he is reputed to have replied, 'He only needs one more "ticket" ' (Challenge Certificate). I do not believe this story to be true, but it is amusing, I think, and we often laugh about it.

Once again, while mentioning shows, I do want to stress the point that to reach the top and to stay there, great perseverance is needed.

Well I remember the occasion when I was showing a lovely bitch who had already won Best-in-Show and two Challenge Certificates, and only required her third Challenge Certificate to become a champion.

The next championship show was in Scotland, so after hours and hours of preparation I set off on the long trip by train. I went up during the day, staying the night at an hotel. The following morning I was up early in time to exercise my dog before the show was due to start. Then off I went hopefully to the show. It often occurs to me, when I see literally hundreds of dogs arriving with their owners and handlers at a show, that they all must be hopeful before judging, or they surely would not be there. How different so many must feel on the way home.

Looking in my catalogue, I found that there were eight entries in the open bitch class. Being entered in the last class of the day, at least I was able to be hopeful for longer than many of the others. My beautiful bitch, looking in the peak of condition, came last in the class.

On the long journey home I had lots of time to think about it all. I find that any set-backs such as this make me more determined than ever to finish the job I have started.

That particular bitch I made a champion soon after this, and she won several more 'tickets' afterwards. So, if you know you have a good one, always remember that it's dogged as does it.

During my long association with dogs, I have made three vows which I keep rigidly. I will tell you why I made them.

The first one was made when I was very young, and with practically no experience in 'doggy business'.

My father had promised to buy me a bitch so that I could breed a litter of puppies. There appeared an advertisement in *Our Dogs* that week of a bitch which, on paper, appeared just what we were looking for. I particularly remember that she was described as having 'good ears, well carried', among other things. My father sent the money off for her and I awaited her arrival with excitement. Imagine my utter disappointment when I opened the box to find that she had completely prick ears. Also, she was covered in 'riff' and looked horrible. While sitting on her box, shedding tears of disappointment, I vowed never to buy another dog without first seeing it, and I have kept this vow for over forty years, so that any mistakes I have made in my purchases have been done with my eyes open.

The second oath I made happened when I had a very bad epidemic of distemper. There were about nine or ten young dogs and bitches in the Kennel at the time. My husband also had a magnificent Beagle with one C.C. to its credit. This was before inoculations were easily obtainable. The most promising four puppies and the hound we took indoors to nurse with electric fires. The remainder were left in a cold corridor kennel. Each had woollen coats, with polo necks, rather like football jerseys.

In spite of injections, which they all had, and very careful and constant nursing for about seven weeks, the four terriers in the house, and the hound, died. Those in the kennel lost a great amount of weight, but they survived.

I then vowed never to keep anything over twelve weeks of age unless it was fully inoculated.

My third vow was made after I heard of friendly business agreements being broken and lifelong friends not speaking to each other!

This vow is, never to have any arrangements or breeding terms with anybody or about any dog.

Finally, I feel that I should give, in concise form, some of the principles that should be apparent from the contents of this book. We cannot all be paragons, but I hope that the novice reader will absorb some of the maxims I suggest, and at least, endeavour to practise the majority. If you do so, I am sure that you will greatly benefit from your hobby and will be justly popular in the doggy fraternity. Needless to say, popularity is a great help to success. I have always been taught that one's chickens come home to roost.

Do *not* criticize other people's dogs—especially in public. Should anyone ask your opinion, give it kindly, remembering how difficult it is to breed a good dog.

Do *not* crowd out your opponent in the show ring.

Do *not* show your dog on a very long lead.

Do *not* use the dog standing next to yours to help to make your own dog show. It is not cricket.

Do *not* use other exhibitors' tables at a show, without first asking their permission.

Do *not* forget that they arrived early, either to erect their own table, or to stake a claim on one provided by the show committee.

Do *not* pass loud remarks when sitting at the show ring.

Do *not* show elation in victory, or depression in defeat.

Do *not* believe the disgruntled exhibitors' criticism of the winners.

Do *not* be rude to the many people, especially at the crowded agricultural shows, who come and say either that they have one just like yours at home, or that breeders are ruining the breed.

Do *not* refuse to accept a reserve prize card (fourth) when there are only four exhibits in the class. If you accept a lower prize card than that which you feel your exhibit deserves, do *not* tear it up for all to see.

Do *not* wear 'rose-coloured glasses' while looking at your own dogs, and thus suffer from 'kennel blindness'.

Do *not* get overstocked; it is the one or two you have in the show ring that matter, not the twenty-five or fifty passengers you have at home.

If you are lucky enough to see, and perhaps be allowed to inspect, a top show specimen, do *not* say that you have a better one at home. Better by far to keep quiet, and one day show it to an astonished world.

Do be patient with everything you do at all times.

Do learn to lose well and to win graciously.

Do accept advice and help in the spirit in which it is given.

Do exercise your exhibit before entering the show ring.

Do brush chalk from your clothing before showing your dog.

Do be tidy yourself. If you cannot keep yourself smart, you certainly cannot make your dog look his best.

Do keep all utensils and travelling boxes scrupulously clean.

Do return travelling boxes promptly, making sure that they are cleaner than when they arrived.

Do remember that a terrier must have good legs and feet.

Do count ten before answering an unkind remark about your dog.

Do be particularly careful not to frighten a young puppy at his first two or three shows.

Do answer all correspondence promptly.

Do remember that you only get out of a dog what you put into it.

Do remember that any one can judge good dogs, but that it takes a good judge to judge bad ones.

Do remember that it is not the length of a pedigree that matters, but its contents.

Do treat dog showing as a sport, rather than as a business.

Do enjoy yourself—it is later than you think!

Needless to say I am constantly breaking some of these rules! However, even to know them is a help, and may, with long experience, result in their application unconsciously. I have learnt everything I know the hard way, and am still learning. It is amazing how new experiences and acquired knowledge constantly arise from the simplest things. I am continually surprised by some little truth, which should have been obvious, and which I had not previously realized.

Personally I do not look upon dogs as work or a nuisance. I thoroughly enjoy every minute I spend with them and count myself fortunate in their company. Whenever I feel at cross-purposes or depressed I remember two of Rudyard Kipling's well-known verses:

The Camel's hump is an ugly lump
Which well you may see at the Zoo;
But uglier yet is the hump we get
from having too little to do.

The cure for this ill is not to sit still,
Or frowst with a book by the fire;
But to take a large hoe and a shovel also,
and dig till you gently perspire.

I then only have to go out into the Kennels and do a bit of pottering about with my dogs, to feel better and happier, and, as they say in Cornwall, 'a new one'. If one has at call such a panacea for one's off times then one must consider oneself very fortunate. I at least, do not intend to give up this advantage, and I have no doubt that I shall enjoy the pleasure of my terriers until the end of my days.

GLOSSARY

Attending the shows, wandering around the benches, and sitting at the ringside, the remarks and expressions of terrier folk must sound very strange to the uninitiated. When I first heard a terrier described as 'a mossy topped one', and another as 'harking a bit', I had not the least idea what was meant. For the sake of similarly puzzled newcomers to the fancy, I give below my interpretation of the more common terms:

Apple-headed Round top of skull

Bred in the purple Of excellent breeding

Bucking about Jumping on the lead, shaking and generally behaving badly

Cackey behind Bad quarters

Cow-hocked Hocks turning inwards

Dolling up Preparing a terrier for show

Gig-lamps Large protruding eyes

Go-away tail Tail carried at a low angle away from the back

Good set-on Tail set high, but not curled, with plenty of posterior behind the tail

Grape-vine (usually) The telephone

Harking a bit Having prick ears

Hound markings Marked like a foxhound, i.e. tan head, with black and tan markings on the body, white predominating

Houndy ears Large, pendulous ears, hanging at the side of the head rather than on the top of the head

Knitting a jumper Crossing front legs when moving

Lacking furnishings Short of face hair and leg hair

Linty coat Soft, open coat on body, face and legs

Loaded in shoulders Muscle on the outside of the upper-arm

Low set Tail right at the end of the back, and the back dropping in front of the tail

Mincing along Not striding out, but rather putting the feet down in very short steps

Mossy topped Soft-coated

Moving like a crab Not moving in a straight line—going sidewards

No heart room Lacks spring of ribs

Out at elbow Turning the outside upper-arm out away from the body

Propping up Holding the lead very tightly under the neck, and pushing the tail straight up while holding the dog in position

Quality The elusive 'something' which no one can describe, but everyone wants. Some call it grace, others lack of coarseness; some say it is outline, some call it presence. Whatever it is called, it cannot be mistaken

Riff A pink skin infection, which can often be seen under the chalk

Shelly body Lack of weight and condition

Spring hocks Walking with the hind legs springing from the ground, as sometimes seen in horses

Spring of ribs The arc of the ribs from the backbone

Straddling Moving with the hind legs too wide apart, and moving in a stiff way, rather than bending the hocks

Straight behind Straight in stifle

Switchback Dip behind shoulders, and uneven topline

Tail like a carrot Tail too straight

Tea-pot handle Curled tail

Tickets Challenge Certificates

Ticks These are the small black spots which often show when a terrier is stripped

Tied in shoulder Often described as two front legs coming out of one hole

Tip-tilted Lower on the front legs than the back ones

Titty faced Thin foreface

Too much daylight Too tall on the legs

Weaving and plaiting Crossing front legs when moving

Well let down behind Hocks low to the ground

Upstanding One that uses its reach of neck well, and stands up elegantly

Could eat in one county and wag his tail in the next Very long in back

Could push a wheelbarrow through him Moving too wide behind. This is every bit as bad a fault as moving too narrow behind

Done a lot of thinking Thick in skull

Would be better on a perch Thin open feet, like a bird's

APPENDIX A

POST-WAR BRITISH WIRE FOX TERRIER CHAMPIONS, 1946 TO 1969

Name	*Sex*	*Sire*	*Dam*	*Owner*	*Breeder*	*Born*
1946						
Ch. Weltona Revelation	D	Culverbrook Tuscan	Hoddlesden Lady	A. Churchill	A. Yates	7.12.43
Ch. Crackley Straightaway	D	Ch. Crackley Supreme Again	Crackley Sequel	J. R. Barlow	Owner	28.6.41
Ch. Kirkmoor Carefree	B	Ch. Miltona Mahmoud	Winkley Show Girl	W. Mitchell	Mrs W. Peel	3.8.41
1947						
Ch. Foxdenton Sunstream	B	Ch. Talavera Jupiter	Barries Vixen	J. N. Hilton	H. Simpson	14.5.43
Ch. Holmwire Hyperion	D	Weltona Axholme Barham	Woodstead Wish	Higginson & Staveley	W. Woods	7.6.44
Ch. Tescot Rita	B	Flexrona Futurist	Tescot Firefly	T. Scott	Owner	23.7.44
Ch. Chief Barmaid	B	Hotel Traveller	Chillie Sauce	R. A. Penny	Owner	1.4.45
Ch. Drakehall Daisybelle	B	Drakehall Debonair	Bedlam Cristabelle	E. Sharpe and A. Butler	E. Sharpe	22.7.45
Ch. Travella Strike	D	Travella Sensation	Travella Gloria	Mrs B. Cole	W. B. Cole	26.9.46
Ch. Tinker Belle	B	Ch. Weycroft Warfare	Princess Tan End	Mrs E. Coton	F. Lowe	2.5.45
Ch. Stocksmoor Sharecroft Select	B	Blue Bird Emperor	Sharefield Surprise	Mrs J. Creasy	J. A. Share	22.3.46
Ch. Crackley Sailaway	D	Crackley Stowaway	Straightlace Susan	J. R. Barlow	Mrs E. G. Robinson	5.5.46

Name	*Sex*	*Sire*	*Dam*	*Owner*	*Breeder*	*Born*
1947—*cont.*						
Ch. Miltona Miss Martha	B	Miltona Magical	Warewell Darkie	W. Hancock	Owner	30.9.42
Ch. Weycroft Woolcomber	D	Ch. Weycroft Warfare	Grendon Success	T. Brampton	O. Wright	25.7.44
Ch. Wyretex Wyns Princess	B	Wyretex Wynstock	Wyretex Wyns Dainty	Mrs D. White	Mr Fullbrook	11.12.43
Ch. Clarington Contender	D	Wireford Colonel	Grenville Countess	W. Seddon	A. Smith	6.6.45
1948						
Ch. Arley Miss Quality	B	Polo Fireaway	Penda Prize Packet	W. Nelmes	Owner	1.8.46
Ch. BlueBird Felicit Fireson	D	Tregolwyn Trademark	Dianes Selected	E. Gray	T. Hill	18.6.46
Ch. Crackley Sunstorm	B	Crackley Satisfaction	Crackley Shelved	T. Ganney	J. R. Barlow	1.11.46
Ch. Glendoune Gavotte	B	Crackley Satisfaction	Crackley Shelved	R. H. McGill	J. R. Barlow	28.4.47
Ch. Casfala Copyright	D	Hotel Traveller	Cawthorne Copyright	F. M. Hughes	J. Roberts	16.6.45
Ch. Cornwall Robecia Radiance	D	Bluebird Emperor	Robecia Streamlined	T. Langley-Jones	T. Goold	20.2.44
Ch. Kirkmoor Anfield Patch-up	B	Ch. Clarington Contender	Anfield Patchwork	M. Seddon	J. Bradley	26.4.47
Ch. Roughsea Sparkle	B	Holmwire Distinction	Deborah of Albernal	H. Holden	J. E. Holden	20.2.45
Ch. Secret Passionette	B	Miltona Magical	Secret Special Session	J. Hudson	Owner	30.8.45
Ch. Talavera Patch-Up	B	Eden Grenadier	Eden Juliet	Mrs H. Graham	F. Robson	28.6.43
Ch. Tornadic Burtona Bonanza Again	D	Burtona Beacon	Burtona Bellona	T. Ganny	C. H. Burton	3.3.45

Ch. Weltona Exelwyre Dustynight	D	Middleforth Tuscan	Juliet of Exelwyre	A. Churchill	J. Yates	13.4.47
Ch. Wynstead What's Wanted	D	Mahmoud Double	Nigels Sunshine	E. Gray	T. Dale	5.11.43
1949 Ch. Beaukrat Ritelsa Rightahead	D	Harrowhill Straightahead	Ritelsa Joy	C. Webb and J. R. Charnley	Miss S. Keenan	14.12.47
Ch. Bedlam Beau Ideal	D	Bedlam Wynstead Woolsack	Bedlam Guda	Mrs A. Butler	A. Butler	24.9.47
Ch. Celtone Constance	B	Ch. Weycroft Woolcomber	Celtone Cobnut	J. Johnson	Owner	16.8.47
Ch. Cockington Cock-a-Hoop	D	Cockington Gay Galliard	Cockington Cover Girl	Mrs M. Cooper	Owner	10.9.47
Ch. Drakehall Miss Banty	B	Axholme Newmaidley Paul	Axholme Delia's Folly	E. Sharpe	G. Burton	7.8.46
Ch. Kirkmoor Cobler	D	Copledene Dante	Kingsbridge Selected	N. Seddon	W. B. Edwards	2.11.47
Ch. Knolbrook Keyman	D	Ch. Clarington Contender	Kingsfield Debutante	J. Greenhalgh	H. Knowles	14.9.47
Ch. Travella Skyliner	D	Ch. Travella Strike	Travella Sunshine	W. Browne-Cole	Owner	7.10.48
Ch. Cornwall Lydbury Lady	B	Ch. Cornwall Robecia Radiance	Crofton Contender	T. Langley-Jones	Mr Mead	8.8.45
Ch. Portbredy Party Piece	B	Penda Pompilius	Penda Prize Packet	Mrs E. L. Williams	Owner	18.10.44
Ch. Gedling Content	B	Gedling Contender	Brewston Patsy	Beresford and Greaves	F. G. Perks	18.10.44
Ch. Drakehall Dairymaid	B	Drakehall Diplomat	Carniff Connie	E. Sharpe	Mr G. Hudson	4.11.47

Name	*Sex*	*Sire*	*Dam*	*Owner*	*Breeder*	*Born*
1949—*cont.*						
Ch. Kebsir Whitwyre Mascot	D	Whitwyre Magnetic	Whitwyre Moti Mahal	G. Hall	Mrs Whitworth	21.3.46
1950						
Ch. Epping Ladycroft Trueform	D	Weltona Realstance	Weltona Ladybird	A. A. W. Simmonds	K. Twyford	14.2.48
Ch. Bedlam Don Juan	D	Bedlam Wynstead Warrant	Nugrade Nublue	Mrs A. Butler	J. W. Holmes	20.8.48
Ch. Casfala Kepple Nobleman	D	Ch. Casfala Copyright	Drakehall de Luxe	F. M. Hughes	A. Barlow	28.2.49
Ch. Winstan Wiswell	D	Ch. Weltona Revelation	Winstan Wendy	E. Winstanley	Owner	19.8.47
Ch. Travella Quick Decision	D	Travella Sensation	Travella Rosebud	W. Browne-Cole	J. J. Bath	14.2.49
Ch. Weltona Realstance	D	Ch. Weltona Revelation	Weltona Carnation	A. Churchill	Owner	20.1.46
Ch. Wycollar Duchess	B	Ch. Crackley Sailaway	Wycollar Wonderous	J. W. Turner	Owner	29.5.49
Ch. Waterton Maryholm Wendy	B	Eden Autocrat	Maryholm Dalfurn Delilah	L. Hastings	A. Clanachan	10.4.48
Ch. Eden Kirkmoor Sunset	B	Ch. Holmwire Hyperion	My Model Miss	F. Robson	F. W. Sutton	20.6.47
Ch. Yours and Mine	B	Stoneycrag Sensation	Lady Yvette	R. F. Harris	H. Cheney	16.5.46
Ch. Burntedge Besum	B	Ch. Weltona Revelation	Talavera Return	J. Hamilton	Owner	20.12.47
Ch. Roundway Rhapsody	B	Roundway Parody	Roundway Rhyming	Mrs J. Creasy	Mrs E. M. Moseley	31.10.47
Ch. Kirkmoor Vanity	B	Ch. Clarington Contender	Wybury Wedlock	Mr and Mrs L. S. Rigby	W. Berry	15.9.47

Ch. Drakehall Delia	B	Drakehall Debonair	Drakehall Dreamgirl	E. Sharpe	Owner	13.2.48
Ch. Polo Fireaway	D	Ch. Crackley Straightaway	Polo Minted	H. W. Hingle	C. B. Cole	21.7.45
1951						
Ch. Drakehall Dandy	D	Danycraig Caich Have a Go	Bedlam Christabelle	H. Harris	E. Sharpe	13.2.50
Ch. Epping Wyrevale Monogram	D	Wyretex Wynstock	Wyrevale Marguerite	A. A. W. Simmonds	W. Miles	29.11.48
Ch. Meyrex Travella Reliable	D	Ch. Travella Strike	Travella Stylist	Miss M. Lloyd	W. Browne-Cole	18.5.49
Ch. Miltona Master Gunner	D	Miltona Marmaduke	Miltona Miss Maud	W. E. Hancock	Owner	13.9.48
Ch. Newmaidley Hob	D	Newmaidley Ceasar	Rexine Fashion	Miss L. G. Beak	F. Hobbs	10.9.48
Ch. Travella Sizzler	D	Ch. Travella Strike	Travella Sunshine	W. Browne-Cole	Owner	19.4.50
Ch. Travella Skyflyer	D	Ch. Travella Strike	Travella Mannequin	W. Browne-Cole	Owner	29.3.49
Ch. Twynstar Accurist	D	Eden Commander	Twynstar Typist	L. W. Edmunds	Owner	23.11.46
Ch. Arley Adorable	B	Ch. Travella Strike	Polo Skylark	W. Nelmes	C. B. Browne-Cole	29.3.50
Ch. Newmaidley Cleopatra	B	Newmaidley Ceasar	Newmaidley Destiny	Miss L. G. Beak	Owner	16.12.46
Ch. Penda Hieover Warrior	B	Penda Pompilius	Hieover Music	Mrs E. Williams	R. Penny	1.4.48
Ch. Ravelly Perfect	B	Crackley Stowaway	Straightlace Susan	R. A. Floyd	Mrs E. Robinson	1.4.47
Ch. Roundway Cats Cradle	B	Ch. Travella Strike	Ch. Stocksmoor Sharecroft Select	Mrs J. Creasy	Owner	1.5.48

Name	Sex	Sire	Dam	Owner	Breeder	Born
1951—*cont.*						
Ch. Summers Silhoutte	B	Summers Security	Foxite Miss	Capt. F. B. Foy	Owner	19.7.49
Ch. Travella Wildcroft Superb	D	Ch. Travella Strike	Littlecroft Hopeful	W. Browne-Cole	E. T. Best	17.12.48
Ch. Torkard Susan	B	Castlecroft Cleanaway	Newmaidley Eve	A. Francis	Owner	31.7.50
Ch. Walldene Wysh	B	Newmaidley Ceasar	Walldene Waltztime	J. Yates	Mrs E. C. Walls	1.4.48
Ch. Westroad Lucky Charm	B	Ch. Cornwall Robecia Radiance	Wyretex Wyns Dream Girl	D. Evans	Owner	15.3.49
Ch. Wyreholme Ember	B	Castlecroft Cleanaway	Wyreholme Entire	D. Williams	Owner	10.3.49
1952						
Ch. Casfala Colonist	D	Ch. Casfala Kepple Nobleman	Casfala Careless	F. M. Hughes	Owner	10.10.50
Ch. Castlecroft Contender Again	D	Castlecroft Cleanaway	Castlecroft Clean Cut	G. A. Prosser	Owner	10.6.49
Ch. Rigwyre Royalist	D	Kirkmoor Connoisseur	Rigwyre Dorincourt Dinah	Mr and Mrs Rigby	Mrs Rigby	23.3.50
Ch. Travella Starshine	D	Ch. Travella Strike	Travella Crystal	W. Browne-Cole	Owner	20.11.51
Ch. Travella Superman	D	Ch. Travella Skyflyer	Travella Carnation	W. Browne-Cole	Owner	12.2.51
Ch. Wyretex Wyns Tuscan	D	Culverbrook Tuscan	Wyretex Wyns Thralia	Mrs D. White	Owner	23.11.45
Ch. Burtona Beseem	B	Castlecroft Cleanaway	Burtona Brimful	C. H. Burton	Owner	14.10.50
Ch. Dinglebank Debutante	B	Ch. Weycroft Woolcomber	Dinglebank Wicked Lady	Mrs M. B. Gardner	Owner	10.1.49

Ch. Graigydan Treacle	B	Ch. Clarington Contender	Graigydan Olga	E. G. Jones	Owner	19.3.49
Ch. Masterlea Lustre	B	Ch. Travella Strike	Lyndoch Lovely Girl	E. G. Bowler	Owner	4.5.51
Ch. Maryholm Mighty Good	D	Ch. Knolbrook Keyman	Fair Pretender	A. Clanachan	W. Haslam	15.8.50
Ch. Burtona Betoken	D	Burtona Bosun	Torkard Countess	C. H. Burton	Owner	14.4.50
Ch. Wyrevale Monomark	D	Ebbw Swell Contender	Wyrevale Marguerite	W. Miles	Owner	31.3.50
Ch. Penda Blackwell Revelation	D	Ch. Weltona Revelation	Edenholme Elfreida	Mrs E. Williams	F. Holliday	12.11.49
Ch. True to Form	B	Ch. Epping Ladycroft Truform	Write Start	Miss B. Cliff	F. Bates	5.8.50
Ch. Newmaidley Dancer	B	Ch. Newmaidley Hob	Newmaidley Kara	Miss L. G. Beak	Owner	24.1.50
Ch. Miltona Miss Conduct	B	Miltona Marmaduke	Miltona Miss Maud	W. Hancock	Owner	13.9.48
Ch. Meyrex Travella Stella	B	Ch. Travella Skyline	Travella Rosebud	Miss M. Lloyd	J. J. Bath	3.10.49
Ch. Castlecroft Contender Again	D	Castlecroft Cleanaway	Castlecroft Cleancut	A. Francis	G. Prosser	10.6.49
1953						
Ch. Quayside Drakehall Duncan	D	Drakehall Ardoch Advocate	Drakehall Dreamgirl	G. E. Rees	E. Sharpe	4.1.52
Ch. Weltona What's This	D	Ch. Wyretex Wyns Tuscan	Full Dress of the Forces	A. Churchill	W. Davies	21.2.52
Ch. Maryholme Northern Monarch	D	Ch. Knolbrook Keyman	Newtown Spitfire	A. Clanachan	S. E. Chapman	20.10.51
Ch. Axholme Double Strike	D	Ch. Travella Strike	Axholme Miss Miranda	G. Burton	Owner	16.8.51

Name	Sex	Sire	Dam	Owner	Breeder	Born
1953—*cont.*						
Ch. Caradochouse Fox Glove	B	Ch. Bedlam Beau Ideal	Crowcroft Mermaid	P. H. Copley	T. Page	12.3.50
Ch. Burntedge Ballet Girl	B	Meritor Moorcrest Mac	Ch. Burntedge Besum	Miss B. Cliff	J. Hamilton	18.8.51
Ch. Tindale Tina	B	Ch. Knolbrook Keyman	Tindale Tay	W. A. Ewbank	Owner	15.12.51
Ch. Travella Allure	B	Ch. Travella Skyflier	Travella Carefree	W. E. Bellerby	W. Browne-Cole	12.9.51
Ch. Milne Reaghcastle Roberta	B	Hartleydene High Flight	Reaghcastle Romance	F. and A. Brooker	C. McCoukey	10.10.51
Ch. Burtona Baliet Girl	B	Ch. Castlecroft Contender Again	Burtona Brunette	C. H. Burton	Owner	12.6.62
Ch. Roundway Strike a Light	D	Ch. Travella Strike	Roundway Wishbone	Mrs. J. Creasy	Owner	24.9.50
1954						
Ch. Karfree Captain	D	Whilwyre Mubarak	Konstallation Kebsir	Mrs. M. Aird	Owner	14.12.47
Ch. Barwyn Welland Wendy	B	Tindale Taft	Welland Wedlock	W. Hancock	L. Jones	1.2.53
Ch. Rawdon Rhoda	B	Ch. Travella Strike	Rawdon Ruby	S. Thorne	W. Andrews	27.12.51
Ch. Crackley Security	D	Crackley Splendid	Crackley Spectant	S. Thorne	J. R. and J. H. Barlow	1.11.52
Ch. Travella Suredo	D	Ch. Travella Starshine	Travella Jasmine	W. Browne-Cole	Owner	11.8.53
Ch. Madam Moonraker	B	Ch. Weltona Exelwyre Dusty Night	Eden Sunshine	J. Stephenson	Owner	12.1.53

Ch. Striking of Laracor	D	Ch. Travella Strike	Gloria of Laracor	Miss A. Hope-Johnstone	Owner	13.11.51
Ch. Wakesgreen Barry	D	Ch. Knolbrook Keyman	Wakesgreen Falstaff Festival	J. D. Sellers	Owner	20.5.51
Ch. Wybury Penda Quicksilver	B	Ch. Penda Blackwell Revelation	Ch. Penda Hieover Warrior	Mrs E. L. Williams	Owner	29.5.51
1955						
Ch. Cawthorne Climax	D	Ch. Burtona Betoken	Cawthorne Twynstar Actionette	Mr and Mrs Pardoe	Owners	26.6.53
Ch. Caradochouse Spruce	D	Drakehall Ardoch Advocate	Caradochouse Rambler Rose	P. H. Copley	Owner	8.2.54
Ch. Travella Sureset	D	Ch. Travella Starshine	Travella Twinkle	W. Browne-Cole	Owner	12.3.54
Ch. Penda Callern Melody	B	Ch. Wyretex Wyns Tuscan	Wyretex Wyns Gloria	Mrs E. L. Williams	E. McCall	1.4.54
Ch. Travella Silk	B	Ch. Travella Starshine	Travella Jasmine	W. Browne-Cole	Owner	11.8.53
Ch. Lemonford Lullaby	B	Rhythmic Remus	Way Ahead	T. Langley-Jones	J. Donaldson	7.7.51
Ch. Lemonford Lollipop of Roundway	B	Ch. Roundway Strike a Light	Ch. Lemonford Lullaby	Mrs J. Creasy	J. Donaldson	30.1.53
Ch. Masterlea Sunspot	B	Ch. Penda Blackwell Revelation	Masterlea Starturn	E. G. Bowler	Owner	2.12.52
Ch. Arley Impressive	B	Ch. Arley Topper	Aversham Lovely Lady	W. Nelmes	G. Ware	1.6.54
Ch. Cawthorne Chloe	B	Ch. Burtona Betoken	Clifton Pandora	Mr and Mrs Pardoe	J. Armitage	16.6.54
Ch. Wyrecliffe Kirkmoor Sunshine	B	Ch. Castlecroft Contender Again	Ardoch Mansbrae Melody	Miss B. Cliff	W. Paul	1.7.52

Name	Sex	Sire	Dam	Owner	Breeder	Born
1955—*cont.*						
Ch. Arley Topper	D	Polo Prince	Polo Skylark	W. G. Nelmes	C. B. Cole	27.1.52
1956						
Ch. Sunnybrook Special Choice	D	Ch. Travella Wildcroft Superb	Sunnybrook Starry Model	Mrs E. Hardy	Owner	12.9.54
Ch. Wyrevale Monotype	D	Ch. Cornwall Robecia Radiance	Wyrevale Marguerite	T. Langley-Jones	W. Miles	17.6.51
Ch. Shoeman's Welland Winston	D	Ch. Axholme Double Strike	Welland Wilwynne	F. Pateman	L. Jones	7.5.54
Ch. Graig-y-Dan Fearnought	D	Ch. Polo Fireaway	Graig-y-Dan Trophy	E. G. Jones	Owner	6.4.52
Ch. Wyretex Wyns Wun-Dar	D	Ch. Wyretex Wyns Tuscan	Ch. Wybury Penda Quicksilver	Mrs D. White	Mrs E. L. Williams	4.5.55
Ch. Whitwyre Field Marshall	D	Ch. Striking of Laracor	Flyagain of the Forces	Mrs Whitworth	Owner	23.10.54
Ch. Penda Peach	B	Ch. Weltona Exelwyre Dusty Night	Ch. Wybury Penda Quicksilver	Mrs E. L. Williams	Owner	19.8.54
Ch. Crackley Standard	D	Crackley Splendid	Crackley Spectant	J. H. and J. R. Barlow	Owners	25.4.54
Ch. Bluebird Ruby of Radwyre	B	Robecia Romulus	Haybarus Harmony	E. Gray	R. W. Southan	14.6.54
Ch. Meritor Say Now	B	Ch. Maryholm Northern Monarch	Moorcrest Model	N. Seddon	Owner	12.4.55
1957						
Ch. Emprise Sensational	D	Ch. Cawthorne Climax	Twynstar Authoress	J. Francis	J. Moss	27.9.55

Ch. Kenelm Supremacy	D	Ch. Burtona Betoken	Barwyn Wildcroft Sunset	J. Bywater	Owner	7.12.55
Ch. Harrowhill Strike Again	D	Ch. Travella Strike	Harrowhill Highlight	Miss Howles	Owner	30.7.52
Ch. Sunnybrook Superjet	D	Ch. Travella Wildcroft Superb	Sunnybrook Starry Model	Mrs E. Hardy	Owner	1.7.56
Ch. Newmaidley Treasure	B	Newmaidley Hannibal	Ch. Newmaidley Dancer	Miss L. G. Beak	Owner	7.8.55
Ch. Lyngarth Social Call	B	Ch. Axholme Double Strike	Lyngarth Serenade	J. Mayfield	Owner	1.7.55
Ch. Wyrecroft Whimsical	B	Wyrecroft Warrior	Wicklewood Twilight	Messrs. Wells & Cartledge	Owners	18.10.55
Ch. Falstaff Forever Amber	B	Ch. Caradochouse Spruce	Mitre Miss Molyneaux	Mrs S. Pinkett	F. and A. Brooker	29.12.55
Ch. Caradochouse Laurel	B	Drakehall Ardoch Advocate	Caradochouse Rambler Rose	P. H. Copley	Owner	8.2.54
Ch. Mitre Miss Spruce	B	Ch. Caradochouse Spruce	Mitre Miss Mavoureen	F. and A. Brooker	Owner	18.1.56
Ch. Florate Fondah	B	Ch. Whitwyre Field Marshall	Florate Fondant	J. H. Smith	Owner	30.9.56
1958 Ch. Anfield Contender	D	Ch. Weltona What's This	Anfield Striking	J. Bradley	J. O'Donnell	4.8.56
Ch. Roundway Bellbhoy	D	Ch. Roundway Strike a Light	Roundway Wedding Belle	Mrs Creasy	Owner	14.10.55
Ch. Gosmore Birthday Boy	D	Ch. Caradochouse Spruce	Gosmore Zeloy Tiara	Mrs A. B. Dallison	Owner	10.5.56

Name	Sex	Sire	Dam	Owner	Breeder	Born
1958—*cont.*						
Ch. Steetonian Skipper	D	Ch. Wyretex Wyns Wundar	Steetonian Suntan	A. Francis	A. G. Dawson	24.1.57
Ch. Crackwyn Corrector	D	Ch. Axholme Double Strike	Mother's Pride	H. Gill	H. Bayles	5.1.57
Ch. Penda Purbeck Repeat	B	Ch. Axholme Double Strike	Tan Lady	Mrs E. L. Williams	J. Sheasby	14.10.46
Ch. Climax Token	B	Ch. Cawthorne Climax	Regal Charmer	M. Crawshaw	L. Sanderson and A Murray	26.6.55
Ch. Windlehurst Susan	B	Ch. Cawthorne Climax	Twynstar Pretty Piece	H. Gill and J. Barlow	J. Moss	1.4.55
Ch. Crackwyn Caprice	B	Ch. Crackley Standard	Crackwyn Cert.	H. L. Gill	Owner	28.3.57
Ch. Flying Alstir High Beamy	B	Ch. Weltona What's This	Lovely Night	Baron Van der Hoop	T. C. Walker	2.6.56
Ch. Kirkmoor Crocus	B	Wyrecliff Sunnybrook Spitfire	Ch. Burntedge Ballet Girl	W. Mitchell	Owner	2.1.56
Ch. Cudhill Kalypso	B	Ch. Wyretex Wyns Wundar	Cudhill Christobel	Dr F. Ogrinz and E. Sharpe	R. Thorpe	2.7.56
1959						
Ch. Seedfield Ardoch Aspiration	D	Al's Barbed Warrior	Ardoch Roundway Model Again	H. M. Harris	J. Clifford	18.4.57
Ch. Crackwyn Captivator	D	Ch. Crackley Standard	Crackwyn Cert.	H. L. Gill	Owner	28.3.57
Ch. Wyretex Wyns Wundarful	D	Ch. Wyretex Wyns Wundar	Purbeck Miss Rimfire	Mrs D. White and D. Stewart	J. Sheasby	25.4.57

Ch. Cornwell Odds On	D	Cornwell Cert.	Ch. Lemonford Lullaby	T. Langley-Jones	Owner	27.9.55
Ch. Zeloy Endeavour	D	Ch. Wyretex Wyns Wundar	Supremacy's Smart Girl	E. Robinson	S. Naylor	21.8.56
Ch. Steelholm Sheena	B	Ch. Burtona Betoken	Clifton Pandora	J. H. Pardoe and E. Sharpe	J. Armitage	23.2.57
Ch. Penda Ravena Snowdrift	B	Shoeman's Pattern	Millbourne Diedre	Mrs E. L. Williams	F. Govier	7.4.57
Ch. Mitre Miss Strike	B	Ch. Gosmore Birthday Boy	Mitre Quicksilver	Mrs B. Jull	F. and A. Brooker	20.5.57
Ch. Penda Oregon Witchcraft	B	Ch. Kenelm Supremacy	Oregon Queen	Mrs E. L. Williams	J. Kirk	30.11.57
Ch. Falstaff Lady Fayre	B	Mitre Advocate	Falstaff Frangrance	Mrs E. Pinkett	Owner	18.12.57
Ch. Crackyn Correct	B	Ch. Crackley Standard	Model Perfect	H. L. Gill	E. G. Lawson	1.6.56
Ch. Penda Cawthorne Cobnut	D	Ch. Cawthorne Climax	Cawthorne Ready Maid	Mrs E. L. Williams	J. Pardoe	9.4.57
Ch. Travella Supercatch	D	Ch. Travella Sureset	Travella Sunflame	W. Browne-Cole	Owner	31.10.56
1960 Ch. Helenstowe Pied Piper	D	Ch. Wyretex Wyns Wundar	Helenstowe Pamela	Mr and Mrs P. Robinson	Owners	23.4.58
Ch. Extreal Realization	D	Extreal Revelation	Crawley Countess	S. Mallam	W. Warburton	30.3.58
Ch. Crackwyn Cockspur	D	Ch. Crackley Standard	Ch. Windlehurst Susan	H. L. Gill	Owner	12.12.58
Ch. Penda Peerless	D	Ch. Penda Cawthorne Cobnut	Ch. Wybury Penda Quicksilver	Mrs E. L. Williams	Owner	27.2.59

Name	Sex	Sire	Dam	Owner	Breeder	Born
1960—*cont.*						
Ch. Whitwyre Even Money	B	Ch. Whitwyre Field Marshall	Whitwyre Maundy Money	Mrs M. Whitworth	Owner	7.11.57
Ch. Crackyn Ardoch Artistic	B	Ch. Anfield Contender	Ardoch Miss Conduct	H. L. Gill	J. Clifford	7.7.58
Ch. Gosmore Arberth Cyclamen	B	Ch. Cornwell Odds On	Arberth Hyacinth	Mrs A. B. Dallison	F. A. Howell	24.9.58
Ch. Weltona Miss Sundance	B	Weycroft Wonderboy	Woldlight Romance	A. Churchill	W. Richmond	11.11.55
Ch. Clennon Chime	B	Ch. Roundway Bellbhoy	Lady Simonetta	Miss N. Fitz-Simons	Owner	11.4.58
Ch. Wicklewood Candybar	B	Ch. Crackley Standard	Wicklewood Flyaway	Miss J. Long	Owner	1.6.58
1961						
Ch. Wyrecliffe Satellite of Senganel	D	Exelwyre Mooroak Aristocrat	Smart Biddy of Senganel	Miss B. Cliff	Mrs A. Smith	16.7.59
Ch. Lyngarth Scout	D	Ch. Zeloy Crusader	Ch. Lyngarth Social Call	J. H. Mayfield	Owner	12.3.60
Ch. Weltona Platta Dainty Princess	B	Ch. Anfield Contender	Mac's Model Wire	A. Churchill	A. Platt	16.5.59
Ch. Kenelm Miss Supremacy	B	Ch. Kenelm Supremacy	Kenelm Gloria	J. Bywater	Owner	27.5.59
Ch. Extreal Elegant	B	Ch. Extreal Realization	Extreal Chorus Girl	S. Mallam	Owner	8.2.60
Ch. Kenelm Conquest	B	Kenelm Odds On	Kenelm Miss Quality	J. Lejeune	J. Bywater	19.2.60

Ch. Seedfield Brooklands Peeress	B	Ch. Seedfield Ardoch Aspiration	Brooklands Coquette	H. M. Harris	H. Johnson	8.5.59
Ch. St. Edmunds Sequel	B	Ch. Cawthorne Climax	Sweet Simonetta	Miss E. Home	Owner	29.6.59
Ch. Oregon Highspot	D	Ch. Kenelm Supremacy	Oregon Queen	J. Kirk	Owner	17.2.59
1962						
Ch. Crackley Cawthorne Compensation	D	Cawthorne Cocoanut	Cawthorne Conquest	Mrs J. Lejeune	Mr A. Tasker	1.4.61
Ch. Zeloy Crusader	D	Ch. Zeloy Endeavour	Zeloy Cinderella	Mr J. Mayfield	Mr E. Robinson	25.9.58
Ch. Gosmore Harwire Hayday	D	Ch. Wyrecliffe Satellite of Senganel	Maltman Sunrise	Mrs A. B. Dallison	Mr Wall	30.5.51
Ch. Taywell Tearaway	D	Ch. Kenelm Supremacy	Taywell Threespire Tabard	Mr H. Powell	Owner	2.12.59
Ch. Zeloy Emperor	D	Ch. Zeloy Endeavour	Zeloy Rhapsody	Mr E. Robinson	Owner	10.3.60
Ch. Penda Daleskirk Caress	B	Ch. Penda Peerless	Purbeck Julie	Mrs E. L. Williams	Mr R. French	5.7.60
Ch. Meritor Baros Jewel	B	Ch. Penda Cawthorne Cobnut	Irish Ch. Baros Wyretex Lilactime	Mr N. Seddon	Mr A. G. Barrett	30.5.60
Ch. Gosmore Mariebel Tina	B	Ch. Zeloy Endeavour	Mitre Miss Marie	Mrs A. B. Dallison	Messrs A. and F. Brooker	5.3.59
Ch. Shoemans Stitcher	B	Kirkmoor Coachman	Shoemans Sciver	Mr F. Pateman	Owner	17.3.61
Ch. Baros Romance	B	Baros Gwenog Tuscan	Baros Storm	Mr A. G. Barrctt	Owner	22.1.60
Ch. Zeloy Moormaides Magic	B	Ch. Zeloy Emperor	Moormaides Cha-Cha-Cha	Mr H. M. Harris	Mr J. Morris	10.5.61

Name	Sex	Sire	Dam	Owner	Breeder	Born
1963						
Ch. Whitwyre Money Market	D	Mitre Advocate	Ch. Whitwyre Even Money	Mrs M. Whitworth	Owner	4.9.61
Ch. Cademans Regent	D	Ch. Axholme Double Strike	Cademans Fashion Queen	Mr E. Massey	Owner	6.6.60
Ch. Weltona Lyngarth Jamboree	D	Ch. Lyngarth Scout	Lyngarth Love Call	Mr A. Churchill	Mr J. Mayfield	3.12.61
Ch. Crackwyn Cock'On	D	Ch. Crackwyn Cockspur	Ch. Crackwyn Ardoch Artistic	Mr H. L. Gill	Owner	30.8.61
Ch. Newmaidley Verdict	D	Newmaidley Barrister	Newmaidley Locket	Miss L. G. Beak	Owner	5.6.59
Ch. Gosmore Empress Sue	B	Ch. Zeloy Emperor	Brooklands Giftie	Mrs A. B. Dallison	Mr W. Dodds	20.8.61
Ch. Gosmore Kirkmoor Tessa	B	Exelwyre Mooroak Aristocrat	Brigston Carosel Miss Fonda	Mrs A. B. Dallison	Mr W. Ratcliffe	9.2.62
Ch. Dunwyre Countess	B	Ch. Axholme Double Strike	Dunwyre Carbonetta	Mr D. H. Carse	Owner	21.4.61
1964						
Ch. Ritelsa Sirius	D	St. Erme Holmwire Simon	Ritelsa Happy Morn	Miss S. Keenan	Owner	25.5.59
Ch. Extreal Replica	D	Ch. Extreal Relization	Extreal Chorus Girl	S. Mallam	Owner	24.9.60
Ch. Penda Tavatina	B	Ch. Penda Peerless	Tavaprim	Mrs E. L. Williams	R. Davison	16.12.61
Ch. Brooklands Elegance	B	Ch. Zeloy Emperor	Brooklands Lola	Miss L. Stella	H. Johnson	29.11.61
Ch. Travella Superstar	D	Travella Starraiser	Travella Serenade	W. Browne-Cole	Owner	25.4.61

Ch. Mitre Dusty Knight	D	Mitre Advocate	Mitre Quicksilver	A. Brooker	Owner	25.2.61
Ch. Wyrecroft War Bonus	D	Ch. Penda Cawthorne Cobnut	Wyrecroft Warpaint	Mrs M. Cartledge	Owner	29.7.62
Ch. Seedfield Ernley Empress	B	Ch. Zeloy Emperor	Townville Trinket	H. M. Harris	C. Whitham	9.5.62
Ch. Wintor Townville Tuscan	D	Townville Traveller	Townville Trinket	Messrs A. and G. Shaw	C. Whitham	10.1.63
Ch. Gosmore Exelwyre Diamond	B	Exelwyre Mooroak Aristocrat	Exelwyre Donatella	V. Mitchell	J. Yates	15.4.63
Ch. Moormaides Mandy	B	Ch. Zeloy Crusader	Moormaides Melody	J. Morris	Owner	8.8.62
Ch. Holmwire Roxville Revision	D	Holmwire Paul Tudor	Roxville Mooremaides Moment	C. H. Higginson	Mr and Mrs W. H. Wright	15.7.63
Ch. Crackyn Connection	D	Ch. Crackley Caw-thorne Compensation	Crackley Spacer	H. L. Gill and J. R. Barlow	H. Skan	23.3.63
1965 Ch. Bengal Ryburn Regent	D	Ch. Zeloy Endeavour	Ryburn Radiance	Mrs Harmsworth	F. H. and F. N. Hopkinson	14.8.63
Ch. Gosmore Kirkmoor Storm	D	Ch. Zeloy Emperor	Model Taste	Mrs A. Dallison	T. Walker	10.10.63
Ch. Wintor Caracus Call Boy	D	Ch. Zeloy Crusader	Nugrade Nesta	A. and G. Shaw	J. Woolley	18.5.64
Ch. Zeloy Select	D	Ch. Zeloy Emperor	Zeloy Tantalizer	E. Robinson	Owner	2.1.62
Ch. Seedfield Conqueror	D	Ch. Zeloy Emperor	Townville Trinket	H. M. Harris	Owner	23.9.63
Ch. Littleway Rose Marie	B	Wakeful White Rajah	Wakeful Rosebud	J. S. Abbott	W. Cobb	29.4.63

Name	Sex	Sire	Dam	Owner	Breeder	Born
1965—*cont.*						
Ch. Meritor Zeloy Sunflower	B	Ch. Zeloy Endeavour	Zeloy Roxville Rainbow	N. Seddon	E. Robinson	2.5.63
Ch. Penda Nugrade Zena	B	Ch. Zeloy Emperor	Nugrade Nena	Mrs E. Williams	J. Holmes	15.9.63
Ch. Platta Smart Susan	B	Ch. Lyngarth Scout	Platta Susan's Princess	W. E. Bellerby	A. Platt	14.3.64
Ch. Rancourt Kirkmoor Cowslip	B	Ch. Zeloy Emperor	Kirkmoor Cygnet	Mrs D. Stewart	Mr and Mrs W. Mitchell	9.9.63
Ch. Sarabel Snapdragon	B	Ch. Crackwyn Cockspur	Ch. Mitre Miss Strike	Mrs B. Jull	Owner	1.11.61
Ch. Worsbro Oladar Royal Maid	B	Cawthorne Contender	Cawthorne Compfrey	F. Robinson	Owner	2.4.62
1966						
Ch. Hatta Boy	D	Ch. Zeloy Emperor	Pride of Main Street	A. S. Booth	Owner	17.10.63
Ch. Weltona Has It	D	Ch. Holmwire Roxville Revision	Ch. Weltona Platta Princess	A. Churchill	Owner	16.1.65
Ch. Wintor Statesman	D	Ch. Wintor Townville Tuscan	Wintor Twilight	A. and G. Shaw	Owners	10.8.64
Ch. Wynecroft Monopoly	D	Wyreworth Justso	Miss Bingo	Mrs M. Cartledge	F. Bates	29.3.64
Ch. Baros Marymount Cinderella	B	Ch. Lyngarth Scout	Baros Delight	A. G. Barrett	L. Pounch	30.6.64
Ch. Gosmore Emprise Elite	B	Emprise Epigram	Peerless Camelia	Mrs A. Dallison	J. Francis	24.2.66

Ch. Gosmore Kirkmoor Content	B	Ch. Zeloy Emperor	Brigston Carosel Miss Fonda	Mrs A. Dallison	Mr and Mrs W. Mitchell	9.9.63
Ch. Nedwar Misslyn	B	Ch. Zeloy Emperor	Miss Delightful	R. Ashworth	T. C. H. Walker	25.9.64
Ch. Sarabel Culswood Chanti	B	Ch. Whitwyre Money Market	Nugrade Nannette	Mrs B. Jull	Mrs M. Cullis	28.4.64
Ch. Wyrecroft Penda Pamela	B	Ch. Zeloy Emperor	Exelwyre Diedre	Mrs M. Cartledge	Mrs T. Smith	5.1.64
1967						
Ch. Gosmore Kirkmoor Satisfaction	D	Ch. Zeloy Emperor	Gosmore Meritor Springtime	Mrs A. Dallison	N. Seddon	24.12.64
Ch. Kirkdale Pirate	D	Exelwyre Mooroak Aristocrat	Mooremaides Merit	G. Higgins	Owner	28.4.65
Ch. Kirkmoor Speculation	D	Holmwire Contender	Platta Star Princess	Mr and Mrs W. Mitchell	A. Platt	1.5.66
Ch. Meritor Sensation	D	Meritor Stringalong	Maryholm Winning Hit	N. Seddon	Owner	11.5.63
Ch. Penda Easelwood Totoplay	D	Ch. Wyrecroft Monopoly	Our Jacynth	Mrs E. L. Williams	Mrs E. Selwood	6.4.66
Ch. Rumsam Rollo	D	Ch. Lyngarth Scout	Cawthorne Catherine	F. Critchley	Owner	29.8.65
Ch. Seedfield Meritor Super Flash	D	Ch. Zeloy Emperor	Maryholm Wintersweet	H. Harris	N. Seddon	8.11.65
Ch. Wyrecroft Woolcomber	D	Ch. Zeloy Crusader	Seedfield Zeloy Zina	Mrs M. Cartledge	T. Brampton	21.2.65
Ch. Wintor Extreal Invader	D	Ch. Extreal Realization	Gussies Girl	S. Mallam	Mrs Westerback	23.12.63
Ch. Gosmore Emprise Elite	B	Emprise Epigram	Peerless Camelia	Mrs A. Dallison	J. Francis	24.2.66

Name	*Sex*	*Sire*	*Dam*	*Owner*	*Breeder*	*Born*
1967—*cont.*						
Ch. Littleway Janaken Viva	B	Ch. Zeloy Emperor	Kirkmoor Cygnet	J. S. Abbott	Mr and Mrs W. Mitchell	8.11.64
Ch. Shoemans Worsbro Wistful	B	Worsbro Wayfarer	Worsbro Whisper	F. Pateman	F. Robinson	9.12.65
Ch. Wintor Express	B	Ch. Wintor Tuscan	Lyngarth True Call	Messrs A. and G. Shaw	Owners	5.4.65
Ch. Wyrecliff Worsbro Whimsical	B	Anfield Betoken	Cawthorne Comfrey	Miss B. Cliff	F. Robinson	25.5.66
1968						
Ch. Penda Worsbro Whistler	D	Worsbro Betoken Again	Ch. Oladar Royal Maid	Mrs E. L. Williams	F. Robinson	1.6.67
Ch. Townville Tally 'O	D	Ch. Wintor Statesman	Townville Teresa	C. Whitham	Owner	4.6.67
Ch. Whitwyre Market Day	D	2nd Ch. Whitwyre Money Market	Whitwyre Miss Elequence	Mrs M. Whitworth	Owner	9.10.64
Ch. Zeloy Exemplar	D	Ch. Zeloy Emperor	Zeloy Ernley Edwina	E. Robinson	Owner	27.6.66
Ch. Zeloy Escort	D	Ch. Zeloy Emperor	Zeloy Tantalizer	E. Robinson	Owner	12.3.64
Ch. Holmwire Tudor Regent	D	Ch. Zeloy Emperor	Holmwire Suntan	W. Prizeman	C. Higginson	2.9.66
Ch. Mooremaides Margo	B	Ch. Zeloy Emperor	Mooremaides Zeloy Corona	J. Morris	Owner	1.6.66
Ch. Sarabel Seasprite	B	Ch. Wyrecroft Woolcomber	Sarabel Silhouette	Mrs B. I. Jull	Owner	27.7.66

Ch. Gosmore Geisha Girl	B	Exelwyre Mooroak Aristocrat	Exelwyre Empress	Mrs A. Dallison	J. Yates	21.9.66
Ch. Penda Ritelsa Silver Spoon	B	Ch. Zeloy Emperor	Ritelsa Radiant	Mrs E. L. Williams	Mrs Heginbotham	16.8.66
Ch. Penda Peppermint	B	Wyrecroft Penda Popular	Ch. Wicklewood Candybar	Mrs E. L. Williams	Mrs M. Cartledge	28.5.67
Ch. Wintor Countess	B	Ch. Wintor Statesman	Wintor Sue	A. and G. Shaw	Owners	8.3.67
1969						
Ch. Gosmore Kirkmoor Craftsman	D	Ch. Kirkmoor Speculation	Kirkmoor Cygnet	Mrs A. Dallison	Mr and Mrs W. Mitchell	23.10.67
Ch. Dimminsdale Crispin	D	Ch. Lyngarth Scout	Dimminsdale Bellona	Mrs Y. Braddock and Mr Nuttall	Mrs Y. Braddock	1.9.67
Ch. Jokyl Wyrecroft Gemini	B	Wyrecroft Penda Popular	Ch. Wicklewood Candybar	Mrs O. Jackson	Mrs M. Cartledge	28.5.67
Ch. Penda Worsbro Weasel	B	Worsbro Betoken Again	Ch. Worsbro Oladar Royal Maid	Mrs E. L. Williams	F. Robinson	1.6.67
Ch. Crispey Nedwar Matilda	B	Ch. Seedfield Meritor Superflash	Ch. Nedwar Miss Lynn	Mr and Mrs W. Havenhand	R. Ashworth	11.11.67
Ch. Littleway Platta Miss Prim	B	Holmwire Contender	Platta Star Princess	J. S. Abbott	A. Platt	1.5.66
Ch. Holmwire Tudor Renown	D	Ch. Zeloy Emperor	Holmwire Suntan	Mr and Mrs Higginson	Owners	2.9.66
Ch. Alkara Mooremaides Bella	B	Ch. Zeloy Emperor	Mooremaides Mayfair	Mr and Mrs Copcull	J. Morris	5.1.68
Ch. Mitre Belle Marie	B	Ch. Holmwire Tudor Regent	Mariebel Countess	A. Brooker	E. Bellerby	16.4.68

APPENDIX B

POST-WAR BRITISH SMOOTH FOX TERRIER CHAMPIONS, 1946 TO 1969

Name	*Sex*	*Sire*	*Dam*	*Owner*	*Breeder*	*Born*
1946						
Ch. Boreham Belsire	D	Boreham Bedad	Boreham Belmalva	The Maharaja of Pithapuram	R. M. Miller	20.10.43
Ch. Lethal Weapon	D	Lethean Waters	Smeatonwood Girlie	L. C. Wilson	T. George	2.7.43
Ch. Leslow Lunette	B	Molton Moonlighter	Danesgate Gwenny	J. Armstrong	Miss E. Smith	1.9.43
Ch. Olton Holdfast	B	Preco Prelude	Ladyship Queen	B. Gradwell	A. Bearsley	4.1.43
1947						
Ch. Abberdale Audacity	D	Abberdale Admiral	Lady Clare	Mr and Mrs Ludford	Mrs E. Wilson	21.1.44
Ch. Flying Fairstead Pilot	D	Stewton Skipper	Fairstead Trixis	Baron Van der Hoop	C. H. Fairs	2.3.45
Ch. Hampole Tinkler	D	Ch. Boreham Belsire	Tidser Tranquil	Miss E. Lindley Wood	Miss E. Sparrow	15.3.46
Ch. Laurel Wreath	D	Ch. Lethal Weapon	Parthings Lassie	L. C. Wilson	G. E. Hurrell	26.2.46
Ch. Twentygrand Avon Joystick	D	Westwood Wizard	Kipyard Miss Marne	N. A. Nicholson	B. Bale	9.10.44
Ch. Cream of Andely	B	Landmark of Andely	Andely Lovely Lady	Mrs B. Lowe-Fallas	Owner	5.7.44
Ch. Darkie Princess	B	Darkie Monarch	Fox Lass	Mr and Mrs E. Mantle	Owners	19.12.43
Ch. Fortuna of Ballig	B	Lethean Waters	Silver Snapper	W. Bateson	Owner	7.10.44
Ch. Wags Dignity	B	Burmar Dan Russel	Zilph	S. Wagstaff	L. Morby	2.11.45

1948						
Ch. Black Andrew	D	Ch. Selecta Rich Reward	Selecta Golden Rule	H. R. Bishop	Owner	30.9.46
Ch. Blybro Spotlight	D	Molten Moonlighter	Santuzza	J. G. Reynolds	W. Newman	17.9.44
Ch. Full Pay	D	Selecta Rich Reward	Normstress	G. Mann	Owner	9.7.46
Ch. Reansway Havanap	D	Saltaire Subaltern	Patchwork	S. C. Flint	Mr and Mrs T. James	16.10.47
Ch. Solus Gold Ore	D	Burmar Dan Russel	Gradley Betty	C. H. Bishop	M. Houghton	27.7.44
Ch. Benmoir Pierette	B	Bournedale Pierrot	Benmoir Exquisite	J. Bendall	Owner	27.12.46
Ch. Blyboro Sunmaid	B	Blyboro Molten Moonseed	Blyboro Melody	N. Bown	I. Guest	16.11.45
Ch. Brooklands Ebony Girl	B	Brooklands Ebony	Glamour Girl of Yealand	H. Johnson	S. Jury	1.10.45
Ch. Farleton Florette	B	Farleton Flyaway	Farleton Fuchsia	Mrs D. R. Richardson	Owner	4.7.47
Ch. Flying Brooklands Venus	B	Wychway Fanfare	Brooklands Queen Bee	Baron Van der Hoop	H. Johnson	13.10.46
Ch. Moorside Mannequin	B	Molton Monarch	Molton Merriment	J. Dobson	W. Bright	13.12.46
Ch. Wright Smart	B	Ch. Travelling Fox	Burmar Bridget	H. M. Harris	M. Houghton	27.7.44
1949						
Ch. Flying Revised Line	D	Ch. Hampole Tinkler	Barrowby Belle	Baron van der Hoop	A Nicholson	1.6.48
Ch. Maryholm Spun Gold	D	Ch. Laurel Wreath	Maryholm Sweetbit	A. Clanachan	Owner	25.1.48
Ch. Selecta Rich Reward	D	Golden Spur of Sker	Lucky Strike	H. R. Bishop	P. Davenport	21.8.43

Name	Sex	Sire	Dam	Owner	Breeder	Born
1949—*cont.*						
Ch. Sorreldene Barrowby Bahram	D	Molton Moonlighter	Barrowby Temptress	Mrs G. W. Bradley	W. Ducker	14.9.46
Ch. Watteau Midas	D	Ch. Laurel Wreath	Brooklands Ebony Belle	Mrs A. Blake	Owner	21.1.48
Ch. Wychway Quintessence	D	Wychway Fanfare	Wychway Delight	W. Shape	Owner	12.10.47
Ch. Brooklands Black Narcissus	B	Wychway Fanfare	Brooklands Queen Bee	H. Johnson	Owner	13.10.46
Ch. Charnworth Sea Storm	B	Ch. Lethal Weapon	Charnworth Liberty Boat	Miss A. D. Cole	Owner	28.6,47
Ch. Chosen Dinah of Notts.	B	Dunedil of Notts.	Cast Iron of Notts.	Her Grace, Kathleen, Duchess of Newcastle	Owner	16.1.48
Ch. Flying Dream	B	Ch. Laurel Wreath	Ch. Flying Brooklands Venus	Baron van der Hoop	Owner	30.7.48
Ch. Hampole Tinkle	B	Ch. Boreham Belsire	Tidser Tranquil	Miss E. Lindley-Wood	Miss E. C. Sparrow	15.3.46
Ch. Lanneau Jewel	B	Firstmonsieur	Lanneau Victoria	J. Lowe	Owner	28.6.47
Ch. Rush Gleam	B	Ch. Lethal Weapon	Rush Modeste	G. Truman-Hewitt, J.P.	Owner	2.4.48
1950						
Ch. Boreham Bendigo	D	Ch. Lethal Weapon	Boreham Between	Dr R. M. Miller	Owner	31.7.47
Ch. Lavish Warpaint	D	Ch. Laurel Wreath	Boreham Bequile	Mrs L. C. Wilson	F. W. Mills	20.2.48

Ch. Maryholme Simon	D	Ch. Maryholme Spun Gold	Maryholme Soncie	A. Clanachan	Owner	4.6.49
Ch. Sheresta Monogram	D	Ch. Selecta Rich Reward	Selecta Golden Rule	H. R. Bishop	Owner	6.1.49
Ch. Touchwood Tribute	D	Ch. Selecta Rich Reward	Wiseclan Coronet	Mrs MacLeod Smith	Owner	1.4.49
Ch. Bourndale Charm	B	Bourndale Rush Recorder	Wintonian Precise	J. S. Smith	W. Bradley	7.5.49
Ch. Farleton Farina	B	Ch. Lethal Weapon	Farleton Fuschia	Mrs D. R. Richardson	Owner	30.3.48
Ch. Kingswood Scroggy Sylphides	B	Clondara Code	Scroggy Princess	A. H. Newbrook	J. Stevenson	15.5.49
Ch. Rosemorder Firefly	B	Ch. Hampole Tinkler	Rosemorder Dainty Lady	F. Furnier	Owner	15.5.48
Ch. Sheresta Miss Andrew	B	Ch. Black Andrew	Emsway Diana	H. R. Bishop	A. W. Wilkinson	14.1.49
Ch. Watteau Lustrous	B	Ch. Full Pay	Watteau Waitress	Mrs A. Blake	Owner	12.12.47
1951 Ch. Brooklands Black Prince	D	Ch. Watteau Midas	Ch. Brooklands Black Narcissus	H. Johnson	Owner	18.10.49
Ch. Farleton Oxhill Bahram	D	Rory of Dowry	Abberdale Prim	Mrs R. D. Richardson	Miss D. J. Beardsley	12.4.50
Ch. On Parade	D	Guns of Victory	Marigold Little Princess	W. Foster	W. McAuley	17.11.49
Ch. Solus Smasher	D	Solus Maryholm Showman	Solus Aprille	C. H. Bishop	Owner	23.4.49

Name	*Sex*	*Sire*	*Dam*	*Owner*	*Breeder*	*Born*
1951—*cont.*						
Ch. Brooklands Happy	B	Ch. Watteau Midas	Brooklands Queen Bee	H. Johnson	Owner	3.10.49
Ch. Flying Larks Wing	B	Ch. Lethal Weapon	Hunston High Jinks	Baron van der Hoop	Mrs K. Southwick	10.6.48
Ch. Hewshott Joyful Light	B	Hewshott Juno	Chosen Light of Notts.	J. F. C. Glover	Owner	10.5.49
Ch. Maryholm Sugar Plum	B	Ch. Maryholm Spun Gold	Benmoir Begonia	A. Clanachan	J. Bendall	29.6.49
1952						
Ch. Beechbank Klesby Heritage	D	Ch. Laurel Wreath	Swisscot Siray Caress	W. Hepwood and Rev. I. D. Knowles	Cap. A. G. Hucklesby	7.5.49
Ch. Brooklands Black Knight	D	Ch. Watteau Midas	Brooklands Queen Bee	H. Johnson	Owner	23.4.51
Ch. Brooklands Lucky Wishbone	D	Ch. Brooklands Black Prince	Burham Birt	H. Johnson	Mrs L. Wilson	22.12.50
Ch. Correct Wartax of Notts.	D	Ch. Lavish Warpaint	Chiffney of Notts.	Her Grace, Kathleen, Duchess of Newcastle	Owner	11.9.49
Ch. Ful-o-Pep	D	Ch. Lavish Warpaint	Pandora's Pet	J. Gough	Owner	16.2.50
Ch. Hampole True Ring	D	Ch. Hampole Tinkler	Hampole Bellrosa	Miss L. Wood	Owner	21.6.49
Ch. Sheresta Model	D	Sheresta Marlbru Marquis	Selecta Golden Rule	H. R. Bishop	Owner	30.5.50
Ch. Broughton Victory Princess	B	Guns of Victory	Marigold Little Princess	Mrs A. G. Boggia	W. McAuley	26.8.50
Ch. Dunold Highland Cream	B	Bourndale Rush Recorder	Dunold Head Barmaid	J. Ellison	Owner	7.9.50

Ch. Hewshott Joyful Lark	B	Hewshott Juro	Chosen Light of Notts.	J. F. C. Glover	Owner	10.5.49
Ch. My Lady of Rivazdal	B	Ch. Hampole Tinkler	Rosemorder Dainty Lady	Mrs H. Haworth	Owner	25.9.50
Ch. Sheresta Borman Vesta	B	Ch. Sheresta Monogram	Borman Victoria	H. R. Bishop	Miss I. Beale	20.8.50
1953						
Ch. Brooklands Black Mask	D	Ch. Brooklands Black Prince	Brooklands Milady	H. Johnson	Owner	1.12.51
Ch. Guisboro Spotlight	D	Ch. Hampole Tinkler	Defiant Jill	D. Holmes	Owner	19.12.50
Ch. Hermon Parthings Loyal Lad	D	Parthings Laddy	Amber Solitaire	Miss K. Emery	Mrs H. Terrell	20.6.51
Ch. Kingswood Kozykole	D	Ch. Lethal Weapon	Charmworth Sea Storm	A. H. Newbrook	Miss A. D. Cole	20.7.52
Ch. Scroggy Sophocles	D	Ch. Lethal Weapon	Scroggy Sultana	J. Stevenson	Owner	15.1.51
Ch. Sheresta Mighty Fine	D	Ch. Sheresta Marlbru Monarch	Sheresta Harmony	H. R. Bishop	Owner	2.2.51
Ch. Stubbington Matinee Idol	D	Ch. Laurel Wreath	Ryden's Danesgate Daisychain	Misses A. Beal and V. B. Sodenberg	Owners	27.4.52
Ch. Charnworth Patched Sails	B	Charnworth Matelot	Ch. Charnworth Sea Storm	Miss A. D. Cole	Owner	18.10.50
Ch. Defray	B	Clondara Code	Stockman's Magnet	P. J. McNamee	Owner	11.8.51
Ch. Gosmore Rosemorder Fireaway	B	Ch. Farleton Oxhill Bahram	Ch. Rosemorder Firefly	Mr and Mrs Dallison	F. Furniss	4.8.52

Name	*Sex*	*Sire*	*Dam*	*Owner*	*Breeder*	*Born*
1953—*cont.*						
Ch. Harkaway Lille	B	Ch. Lethal Weapon	Harkaway Likeable	Miss B. Stapley	Owner	2.9.50
Ch. Watteau Songstress	B	Ch. Lavish Warpaint	Wildflower	Mrs A. Blake	D. Kay	24.4.51
1954						
Ch. Brooklands Black Ace	D	Ch. Brooklands Lucky Wishbone	Brooklands Black Tulip	H. Johnson	Owner	9.11.52
Ch. Hampole Housemaster	D	Ch. Hampole Tumbler	Hampole Homechat	Miss E. L. Wood	Owner	29.5.53
Ch. Lanneau Jerrod	D	Lanneau Hayespark Topnotcher	Kentucky Teddibar Aileen	J. Lowe	E. Barnard	22.10.52
Ch. Rush Pegasus	D	Ch. Watteau Midas	Rush Modeste	G.Truman-Hewitt,J.P.	Owner	23.12.52
Ch. Barwyn Princess	B	Bournedale Rush Recorder	Blyboro Barley	T. F. Lewin	Owner	14.8.52
Ch. Burmar Dawn	B	Burmar Lance	Burmar Adriana	Mrs E. Marshall and Miss E. G. Burton	Owners	9.8.52
Ch. Farleton Gay Florentina	B	Ch. Farleton Oxhill Bahram	Ch. Farleton Florette	Mrs D. Roy Richardson	Owner	8.3.53
1955						
Ch. Brooklands Lucky Dip	D	Ch. Brooklands Lucky Wishbone	Ch. Brooklands Black Narcissus	H. Johnson	Owner	28.9.53
Ch. Hewshott Jaguar	D	Hewshott Javelin	Hewshott Joy's Wish	J. F. C. Glover	Owner	13.5.53
Ch. Oxhill Band Leader	D	Farleton Fine Fellow	Oxhill Bridesmaid	A. Baldwin	Owner	8.6.53
Ch. Samarium Jan	D	Ch. Beechbank Klesby Heritage	Princess Priceless	Miss A. K. Keats and J. Cropper	J. Cropper	20.5.53

Ch. Wychway Buccaneer	D	Wychway Newsboy	Wychway Tinker Belle	W. Snape	J. L. Gee	12.2.53
Ch. Farleton Amber Gaiety Girl	B	Ch. Farleton Oxhill Bahram	Thealelands Jill	Mrs D. Roy Richardson	J. P. Miller	13.12.53
Ch. Hermon Palmist	B	Ch. Hermon Parthings Loyal Lad	Crystal Lady	Miss K. Emery	Owner	18.7.54
Ch. Kenelm Bellechien Pirouitte	B	Ch. Lethal Weapon	Bellechien Ballet Girl	J. W. Bywater	Mr and Mrs R. Kiesekoms	4.4.53
Ch. Lady Fox of Tutbury	B	Ch. Sheresta Monogram	Cannily of Notts.	C. E. Ballance	Owner	20.9.51
Ch. Maryholm Sweetmeat	B	Ch. Farleton Oxhill Bahram	Maryholme Sweetbit	A. Clanachan	Owner	24.3.54
Ch. Wenn of Ballig	B	Ingham Racket	Beltuna of Ballig	W. Bateson	Owner	20.5.53
1956						
Ch. Burmar Warrior	D	Burmar Lance	Burmar Adriana	Mrs E. L. Marshall and Miss S. G. Burton	Owners	21.11.53
Ch. Farleton Barrowby Barney	D	Ch. Oxhill Bandleader	Lady Jewel	Mrs D. R. Richardson	W. Ducker	26.6.54
Ch. Lanneau Jeremy	D	Ch. Lanneau Jerod	Lanneau Jenta	Mrs and Mr J. Lowe	J. Lowe	20.12.54
Ch. Solus Rosemorder Fire Alarm	D	Ch. Hermon Parthings Loyal Lad	Rosemorder Yes Milady	C. H. Bishop	F. Furniss	24.9.54
Ch. Watteau Chorister	D	Ch. Brooklands Lucky Wishbone	Ch. Watteau Songstress	Mrs A. Blake	Owner	3.11.54
Ch. Hampole Housewife	B	Hampole Campsmount Warbond	Hampole Homechat	Miss E. Lindley Wood	Owner	22.5.54

Name	*Sex*	*Sire*	*Dam*	*Owner*	*Breeder*	*Born*
1956—*cont.*						
Ch. Harkaway Eliza	B	Ch. Hermon Parthings Loyal Lad	Harkaway Lilli	Miss B. Stapley	Owner	24.8.54
Ch. Parkend Druscilla	B	Parkend Director	Parkend Diana	N. Bown	Owner	17.3.54
1957						
Ch. Hewshott Juggler	D	Ch. Hewshott Jaguar	Hewshott Jigsaw	J. F. C. Glover	Owner	5.3.56
Ch. Last o' Weapon	D	Ch. Lethal Weapon	Lesser Waxbill	Lynn C. Wilson	Mrs L. C. Wilson	31.8.55
Ch. Maryholm Sailaway	D	Maryholm St Patrick	Maryholm So Sweet	A. Clanachan	Owner	15.5.55
Ch. Parkend Delegate	D	Parkend Director	Parkend Diana	N. Bown	Owner	8.2.56
Ch. Parkend Director Again	D	Parkend Director	Parkend Diana	N. Bown	Owner	25.6.55
Ch. Charnworth Sea Nymph	B	Ch. Kingswood Kozy Kole	Ch. Charnworth Patched Sails	Miss A. D. Cole	Owner	14.12.53
Ch. Farleton Saltholme Saucy	B	Ch. Rush Pegasus	Saltholme Sally	Mrs D. Roy Richardson	A. M. Simpson	31.7.55
Ch. Harkaway Emma	B	Ch. Hermon Parthings Loyal Lad	Ch. Harkaway Lilli	Mrs E. L. Marshal and Miss E. G. Burton	Miss Stapley	24.8.54
Ch. Maryholme Silver Lady	B	Ch. Maryholme Sailaway	Maryholme Silver Mist	A. Clanachan	Owner	7.6.56
Ch. Parthings Land Girl	B	Parthings Laddie	Amber Solitaire	G. E. Hurrell	Mrs H. Terrell	4.4.53
Ch. Shaftmoor Sundew	B	Farleton Fine Fellow	Platoon Commander	A. Lloyd	E. Sandland	31.8.56
1958						
Ch. Auchencrosh Jack Boot	D	Hunt Master	Auchencrosh Autigons	Mr and Mrs H. J. Hewetson	W. Platten	28.8.56

N

Ch. Brooklands Royaltan	D	Brooklands Decorator	Brooklands Sparkle	H. Johnson	Owner	25.12.56
Ch. Charneth Choir Boy	D	Charneth Call Boy	Charneth Gipsy	C. K. Boden	Owner	7.4.57
Ch. Glascoed Gangster	D	Ch. Wychway Buccaneer	Glascoed Garland	S. A. Wheeler	J. J. Lynch	4.5.55
Ch. Hermon Rebel	D	Ch. Hermon Parthings Loyal Lad	Crystal Lady	Miss K. Emery	Owner	1.10.56
Ch. Solus Marilyn	B	Ch. Solus Rosemorder Fire Alarm	Wraysdale Enterprise	C. H. Bishop	Mrs V. Gold	10.10.56
Ch. Watteau Sonata	B	Ch. Watteau Chorister	Watteau Marylyn	Mrs A. Blake	Owner	16.3.57
1959						
Ch. Brooklands Happy Wish	D	Ch. Brooklands Lucky Wishbone	Ch. Brooklands Happy	H. Johnson	Owner	10.10.56
Ch. Glascoed Guinea Gold	D	Ch. Maryholm Spun Gold	Glascoed Glitter	S .A. Wheeler	J. J. Lynch	22.6.57
Ch. Maryholm Sureline	D	Laurel of Din	Ch. Maryholm Sweetmeat	A. Clanachan	Owner	22.4.58
Ch. Watteau Madrigal	D	Ch. Watteau Chorister	Watteau Marylyn	Mrs A. Blake	Owner	4.5.58
Ch. Watteau Merry Thought	D	Ch. Brooklands Happy Wish	Watteau Skylark	Mrs A. Blake	Owner	27.9.57
Ch. Burmar Snocat	B	Ch. Burmar Warrior	Ch. Harkaway Emma	Mrs E. L. Marshall and Miss E. G. A. Burton	Owners	11.2.58
Ch. Forthill Fascination	B	Barrack Bandlader	Molton Melia	J. Smyth	Owner	16.1.58
Ch. Hampole Fidelity	B	Ch. Hermon Parthings Loyal Lad	Ch. Hampole Housewife	Miss E. Lindley Wood	Owner	15.12.57

Name	Sex	Sire	Dam	Owner	Breeder	Born
1959—*cont.*						
Ch. Lanneau Jessica	B	Ch. Brooklands Black Ace	Lanneau Jenta	Mr and Mrs J. Lowe	Owners	10.4.56
Ch. Silver Mannequin	B	Laurel of Din	Joans Gem	J. Magill	Owner	3.9.57
Ch. Watteau Rhapsody	B	Ch. Watteau Chorister	Watteau Marylyn	Madam M. Soudee	Mrs A. Blake	4.5.58
1960						
Ch. Solus Soloist	D	Ch. Watteau Chorister	Ch. Solus Marilyn	C. H. Bishop	Owner	11.2.59
Ch. Hermon Card Trick	B	Ch. Burmar Warrior	Ch. Hermon Palmist	Miss K. Emery	Owner	15.11.58
1961						
Ch. Ellastone Gold Nugget	D	Ch. Watteau Chorister	Brooklands Lady Alice	K. Dickenson	Owner	15.12.59
Ch. Foremark Festive	D	Brooklands Decorator	Watteau Skylark	Mrs W. Newbury	Mrs A. Blake	21.12.59
Ch. Maryholm Nornay Mainsail	D	Ch. Maryholm Sailaway	Maryholm Shamrock	A. Clanachan	Mesdames M. Coward and F. M. Soubrey	15.10.59
Ch. Brooklands Spice	B	Ch. Watteau Chorister	Brooklands Joybelle	H. Johnson	Owner	12.7.60
Ch. Flying Hermon Diamond	B	Ch. Hermon Parthings Loyal Lad	Crystal Lady	Baron van der Hoop	Miss K. Emery	29.4.59
Ch. Watteau Cantata	B	Ch. Watteau Chorister	Watteau Marylyn	Mrs A. Blake	Owner	16.5.59
1962						
Ch. Maryholm Ship Ahoy	D	Maryholm Nornay Mainsail	Ch. Maryholm Silver Lady	A. Clanachan	Owner	9.2.61
Ch. Lanneau Jekyll	D	Lanneau Jeweller	Lanneau Jezebel	Mr and Mrs J. Lowe	Owners	18.11.60

Ch. Mattocline	D	Ch. Maryholme Sureline	Keyworth Pagegirl	P. H. Ireson	Owner	2.9.59
Ch. Viaduct Beau Brummel	D	Viaduct Golden Rod	Viaduct Carmen	Miss S. Langstaff and E. Lockey	Owners	12.6.59
Ch. Newmaidley Jehu	D	Ch. Brooklands Lucky Wishbone	Newmaidley Destiny	Miss L. G. Beak	Owner	8.5.60
Ch. Burmar Emily	B	Burmar Arrow	Ch. Harkaway Emma	Mrs Marshall and Miss Burton	Owners	1.8.60
Ch. Hermon Fantasy	B	Ch. Hermon Rebel	Hermon Witchcraft	Miss K. Emery	Lady Gooch	14.5.60
Ch. Hermon Blacklands Sophia	B	Ch. Solus Soloist	Blacklands Jane	Miss K. Emery	Lt-Col and Mrs D. Yate-Lee	22.11.60
Ch. Nornay Topsail	B	Ch. Maryholm Sailaway	Maryholm Shamrock	Mesdames M. Coward and F. M. Soubrey	Owners	24.11.60
Ch. Maryholm Royal Surepay	B	Roylan Rich Reward	Roylan Golden Gem	H. Thomas	J. Morton	27.5.59
1963 Ch. Watteau Snuff Box	D	Watteau Sculpture	Beechbank Olive	Mrs A. Blake	Owner	1.2.62
Ch. Lesterley Starliner	D	Ch. Mattocline	Hermon Stargazer	Miss M. E. Lambert	Owner	16.2.61
Ch. Solus Sidewater Seahawk	D	Ch. Watteau Chorister	Sidewater Snow Fairy	C. H. Bishop	Mrs V. Goold	18.10.60
Ch. Brooklands Present	B	Ch. Watteau Chorister	Watteau Marylyn	H. Johnson	Mrs A. Blake	16.5.59
Ch. Astonabbotts Fair Dinkum	B	Burmar Major George	Aston Sally	P. Kempster	Owner	11.6.60
Ch. Maryholm Silver Snowflake	B	Ch. Maryholm Sureline	Ch. Maryholm Silver Lady	A. Clanachan	Owner	1.9.59

Name	*Sex*	*Sire*	*Dam*	*Owner*	*Breeder*	*Born*
1963—*cont.*						
Ch. Hewshott Jennie Jerome	B	Hewshott Jerome	Hewshott Jane Eyre	J. F. C. Glover	Owner	16.2.62
Ch. Harkaway Holly	B	Ch. Solus Soloist	Harkaway Dinah-Mite	Miss B. Stapley	Owner	5.12.60
1964						
Ch. Newmaidley Vodka	D	Ch. Watteau Snuff Box	Newmaidley Destiny	Mrs L. G. Beak	Owner	18.1.63
Ch. Shaftmoor Bellechien White Heather	B	Ch. Hermon Rebel	Bellechien Songstress	A. Lloyd	Mr and Mrs Kiesekoms	24.9.62
Ch. Spaceman	D	Debough	Wynnor's Fancy	J. Russell	N. Simpson	14.5.61
Ch. Casterbridge Starlight	B	Ch. Lesterley Starliner	Casterbridge Fairy Snow	C. N. Rippingale	Owner	28.10.63
Ch. Greenbelt Coppernob	D	Lanneau Jerome	Greenbelt Autumn Tint	Mrs L. Brady	Owner	20.4.61
Ch. Conformable Benjamin	D	Ch. Glascoed Guinea Gold	Conformable Sexta	S. A. Wheeler	Owner	23.6.62
Ch. Hermon Snow White	B	Hiya Harbourmaster	Hiya Hanwen	Miss K. Emery	Mrs A. M. Tomlinson	28.6.62
1965						
Ch. Ellastone Lucky Nugget	D	Ch. Ellastone Gold Nugget	Ellastone Lucky Dip	K. Dickinson	Owner	7.6.62
Ch. Foremark Ebony Box	D	Ch. Watteau Snuff Box	Watteau Gaybird	Mrs W. Newbury	Owner	9.10.63
Ch. Harkaway Lancashire Lad	D	Ch. Watteau Snuff Box	Harkaway Mandy	Miss B. Stapley	Owner	26.1.63

Ch. Newmaidley Joshua	D	Ch. Watteau Snuff Box	Newmaidley Kala	Miss L. Beak	Owner	9.10.63
Ch. Hampole Hero	B	Ch. Hermon Parthings Loyal Lad	Ch. Hampole Housewife	B. Walker	Miss E. Lindley-Wood	30.6.59
Ch. Hermon Mirage	B	Ch. Watteau Snuff Box	Ch. Hermon Fantasy	Miss K. Emery	Owner	24.8.63
1966 Ch. Fernery Grand Duke	D	Fernery Viscount	Honey Sweet of Far Green	T. Fennyhough	Owner	15.10.61
Ch. Lanneau Jevron	D	Lanneau Jepcot	Lanneau Jeminy	Mr and Mrs J. Lowe	Owners	10.6.65
Ch. Parkend Democrat	D	Ch. Parkend Director Again	Parkend Donna	N. Bown	Owner	19.9.64
Ch. Viscum Voluntary	D	Hampole Tatler	Ch. Hampole Hero	B. Walker	Owner	30.3.64
Ch. Foxformee Spots	B	Ch. Harkaway Lancashire Lad	Foxformee Burmar Unity	Mrs M. Greenslade	Owner	18.2.65
Ch Lanneau Jessie	B	Ch. Watteau Madrigal	Lanneau Jezebel	Mr and Mrs J. Lowe	Owners	11.1.64
Ch. Maryholm Gelston Impala	B	Ch. Maryholm Sureline	Solus Tanis	A. Clanachan	The Hon. A. Oakshott	22.10.64
1967 Ch. Ellastone Firecrest	D	Ch. Ellastone Gold Nugget	Marteg Maybird	K. Dickenson	Mr and Mrs E. R. Davies	30.6.65
Ch. Lesterley Starliner	D	Ch. Mattocline	Hermon Stargazer	Miss M. E. Lambert	Owner	16.2.61
Ch. Foxformee Adelaide	B	Ch. Harkaway Lancashire Lad	Foxformee Burmar Unity	Miss M. Bagot	Mrs M. Greenslade	27.3.66

Name	*Sex*	*Sire*	*Dam*	*Owner*	*Breeder*	*Born*
1967—*cont.*						
Ch. Hermon Snow Drift	B	Ch. Harkaway Lancashire Lad	Ch. Hermon Snow White	Miss K. Emery	Owner	16.6.65
Ch. Watteau Last Word	D	Watteau S'nufsed	Lingrove Linnet	Mrs and Miss A. Blake	Owners	4.4.66
Ch. Hermon Snow Fall	B	Ch. Harkaway Lancashire Lad	Ch. Hermon Snow White	Miss K. Emery	Owner	16.6.65
Ch. Maryholm Siller Belle	B	Ch. Maryholm Sureline	Ch. Maryholm Silver Lady	A. Clanachan	Owner	31.10.65
Ch. Nornay Windward	B	Nornay Conviction	Ch. Nornay Topsail	Mrs M. Coward and Mrs F. M. Soubry	Owners	5.8.64
1968						
Ch. Nornay Navigator	D	Nornay Conviction	Ch. Nornay Topsail	Mesdames Coward and Soubry	Owners	6.1.67
Ch. Lanneau Jethro	D	Ch. Lanneau Jevron	Ch. Lanneau Jessie	Mr and Mrs J. Lowe	Owners	3.8.67
Ch. Herman Snowman	D	Ch. Harkaway Lancashire Lad	Ch. Harmon Snow White	Miss K. Emery	Owner	31.10.66
Ch. Davesfame Kompliment	B	Burmar Major George	Hampole Chitchat	Master David Kitchen	D. Kitchen	1.7.66
Ch. Watteau Happy Talk	B	Watteau S'nufsed	Lingrove Linnet	Mrs and Miss A. Blake	Owners	4.4.66
Ch. Herman Snowball	B	Ch. Harkaway Lancashire Lad	Ch. Hermon Snow White	Miss K. Emery	Owner	31.10.66

Ch. Newmaidley Naomi	B	Ch. Newmaidley Vodka	Newmaidley Kala	Miss L. Beak	Owner	20.3.65
Ch. Spritely Chorus Girl	B	Ch. Newmaidley Vodka	Spritely Lady	A. Thomson	Miss Robinson	12.12.64
1969						
Ch. Barrowby Martha	B	Barrowby Consul	Barrowby Daphne	W. Ducker	Owner	22.6.67
Ch. Thermfare Inca	B	Thermfare Elegance	Andersley Abbess	J. Pitcairn	Owner	23.1.68
Ch. Hermon Snowflake	B	Ch. Harkaway Lancashire Lad	Ch. Hermon Snow White	Miss K. Emery	Owner	23.1.68
Ch. Newmaidley Cossack	D	Ch. Newmaidley Vodka	Newmaidley Leira	Miss L. Beak	Owner	7.12.66

APPENDIX C

KENNEL CLUB REGISTRATIONS

	Wire Fox Terriers	*Smooth Fox Terriers*
1946	7,986	2,228
1947	8,227	2,338
1948	6,603	1,879
1949	5,779	1,825
1950	4,674	1,787
1951	4,220	1,640
1952	3,157	1,211
1953	2,566	1,079
1954	2,283	1,077
1955	2,406	1,085
1956	2,358	1,088
1957	2,178	926
1958	2,223	866
1959	2,455	951
1960	2,503	965
1961	2,458	821
1962	2,491	900
1963	2,462	770
1964	2,279	758
1965	2,255	723
1966	1,786	609
1967	1,650	607
1968	1,612	591

APPENDIX D

FOX TERRIER CLUBS

Birmingham and District Fox Terrier Club—J. W. Bywater, Woodbine Cottage, Coventry Road, Bickenhill, Solihull, Warwicks.

Bolton and District Fox Terrier Club—C. A. Williams, 59, Chetwyn Avenue, Bromley Cross, Bolton, Lancs.

Fox Terrier Club—Miss E. Swyer, The Pigeons, Bramley Green, Basingstoke, Hants.

Fox Terrier Club of Scotland—J. S. Abbott, Whitehill of Coodham, Symington, Ayrshire.

Mansfield Woodhouse Fox Terrier Club—Mrs J. Painter, 121, Westfield Lane, Mansfield, Notts.

Northern Ireland Fox Terrier Club—H. Jackson, 109, Duneden Park, Belfast 14, Northern Ireland.

Northumberland and Durham Fox Terrier Club—G. K. Glister, 17, Railway Street, Langley Moor, Co. Durham.

Notts. Fox Terrier Club—Mrs D. J. Daly, 'Sunnydene', Kirby Bellars, near Melton Mowbray, Leicester.

Oldham Fox Terrier Club—F. Furniss, 125, Greyfriars Court, St. Simon Street, Salford 3, Lancs.

Sheffield Fox Terrier Club—Mrs M. Cook, 79, Studmoor Road, Kimberworth Park, Rotherham, Yorkshire.

Smooth Fox Terrier Association—Miss B. Gough, Hele Cottage, Ashburton, South Devon.

South Durham and North Yorkshire Fox Terrier Club—Mr W. Taylor, 128, Brougham Street, Darlington.

Sunderland Fox Terrier Club—D. G. Storey, 2, Depot Cottages, Spennymoor, Co. Durham.

Ulster Fox Terrier Club—Mr G. McCann, 35, Breda Park, Belfast, 8.

Wire Fox Terrier Association—G. M. P. Williams, O.B.E., 37, Avenue Road, St. Albans, Herts.

Yorkshire Fox Terrier Association—E. Hardy, 'Hill Crest', Wakefield Road, Hemsworth, Near Pontefract, Yorks.

APPENDIX E

WHELPING TABLE

TABLE SHOWING WHEN A BITCH IS DUE TO WHELP

Served January	Due to Whelp March	Served February	Due to Whelp April	Served March	Due to Whelp May	Served April	Due to Whelp June	Served May	Due to Whelp July	Served June	Due to Whelp August	Served July	Due to Whelp September	Served August	Due to Whelp October	Served September	Due to Whelp November	Served October	Due to Whelp December	Served November	Due to Whelp January	Served December	Due to Whelp February
1	5	1	5	1	3	1	3	1	3	1	3	1	2	1	3	1	3	1	3	1	3	1	2
2	6	2	6	2	4	2	4	2	4	2	4	2	3	2	4	2	4	2	4	2	4	2	3
3	7	3	7	3	5	3	5	3	5	3	5	3	4	3	5	3	5	3	5	3	5	3	4
4	8	4	8	4	6	4	6	4	6	4	6	4	5	4	6	4	6	4	6	4	6	4	5
5	9	5	9	5	7	5	7	5	7	5	7	5	6	5	7	5	7	5	7	5	7	5	6
6	10	6	10	6	8	6	8	6	8	6	8	6	7	6	8	6	8	6	8	6	8	6	7
7	11	7	11	7	9	7	9	7	9	7	9	7	8	7	9	7	9	7	9	7	9	7	8
8	12	8	12	8	10	8	10	8	10	8	10	8	9	8	10	8	10	8	10	8	10	8	9
9	13	9	13	9	11	9	11	9	11	9	11	9	10	9	11	9	11	9	11	9	11	9	10
10	14	10	14	10	12	10	12	10	12	10	12	10	11	10	12	10	12	10	12	10	12	10	11
11	15	11	15	11	13	11	13	11	13	11	13	11	12	11	13	11	13	11	13	11	13	11	12
12	16	12	16	12	14	12	14	12	14	12	14	12	13	12	14	12	14	12	14	12	14	12	13
13	17	13	17	13	15	13	15	13	15	13	15	13	14	13	15	13	15	13	15	13	15	13	14
14	18	14	18	14	16	14	16	14	16	14	16	14	15	14	16	14	16	14	16	14	16	14	15
15	19	15	19	15	17	15	17	15	17	15	17	15	16	15	17	15	17	15	17	15	17	15	16
16	20	16	20	16	18	16	18	16	18	16	18	16	17	16	18	16	18	16	18	16	18	16	17
17	21	17	21	17	19	17	19	17	19	17	19	17	18	17	19	17	19	17	19	17	19	17	18
18	22	18	22	18	20	18	20	18	20	18	20	18	19	18	20	18	20	18	20	18	20	18	19
19	23	19	23	19	21	19	21	19	21	19	21	19	20	19	21	19	21	19	21	19	21	19	20
20	24	20	24	20	22	20	22	20	22	20	22	20	21	20	22	20	22	20	22	20	22	20	21
21	25	21	25	21	23	21	23	21	23	21	23	21	22	21	23	21	23	21	23	21	23	21	22
22	26	22	26	22	24	22	24	22	24	22	24	22	23	22	24	22	24	22	24	22	24	22	23
23	27	23	27	23	25	23	25	23	25	23	25	23	24	23	25	23	25	23	25	23	25	23	24
24	28	24	28	24	26	24	26	24	26	24	26	24	25	24	26	24	26	24	26	24	26	24	25
25	29	25	29	25	27	25	27	25	27	25	27	25	26	25	27	25	27	25	27	25	27	25	26
26	30	26	30	26	28	26	28	26	28	26	28	26	27	26	28	26	28	26	28	26	28	26	27
			MAY																				
27	31	27	1	27	29	27	29	27	29	27	29	27	28	27	29	27	29	27	29	27	29	27	28
	AP	R.																				MAR	
28	1	28	2	28	30	28	30	28	30	28	30	28	29	28	30	28	30	28	30	28	30	28	1
						JU	LY																
29	2	29	3	29	31	29	1	29	31	29	31	29	30	29	31	29	31	29	31	29	31	29	2
					JU	NE			AUG.		SE	P.	OCT.		NOV.		DEC.		JAN.		FEB.		
30	3			30	1	30	2	30	1	30	1	30	1	30	1	30	2	30	1	30	1	30	4
31	4			31	2			31	2			31	2	31	2			31	2			31	3

INDEX